Lecture Notes in Mathematics   1676

Editors:
A. Dold, Heidelberg
F. Takens, Groningen

Springer
Berlin
Heidelberg
New York
Barcelona
Budapest
Hong Kong
London
Milan
Paris
Santa Clara
Singapore
Tokyo

Pilar Cembranos   José Mendoza

Banach Spaces of
Vector-Valued Functions

 Springer

Authors

Pilar Cembranos
José Mendoza
Departamento de Análisis Matemático
Facultad de Ciencias Matemáticas
Universidad Complutense de Madrid
E-28040 Madrid, Spain
e-mail: cembrp@sunam1.mat.ucm.es
          mendoza@sunam1.mat.ucm.es

Cataloging-in-Publication Data applied for
Die Deutsche Bibliothek - CIP-Einheitsaufnahme

Cembranos, Pilar:
Banach spaces of vector valued functions / Pilar Cembranos ; José
Mendoza. - Berlin ; Heidelberg ; New York ; Barcelona ; Budapest ;
Hong Kong ; London ; Milan ; Paris ; Santa Clara ; Singapore ;
Tokyo : Springer, 1997
  (Lecture notes in mathematics ; 1676)
  ISBN 3-540-63745-1

Mathematics Subject Classification (1991): 46E40, 46B20, 28B05

ISSN 0075-8434
ISBN 3-540-63745-1 Springer-Verlag Berlin Heidelberg New York

© Springer-Verlag Berlin Heidelberg 1997
Printed in Germany

The use of general descriptive names, registered names, trademarks, etc. in this
publication does not imply, even in the absence of a specific statement, that such
names are exempt from the relevant protective laws and regulations and therefore
free for general use.

Typesetting: Camera-ready T_EX output by the authors
SPIN: 10553403   46/3142-543210 - Printed on acid-free paper

For Emilio, Dolores, Luis, Javier and Pablo Mendoza Cembranos

Table of Contents

Introduction

The study of vector-valued function Banach spaces is a very active field of research, and it is already very old, too. It begins soon after Banach's book, in the thirties and early forties, with the classical works of Birkhoff, Boas, Bochner, Clarkson, Day, Dunford, Gelfand, Pettis, Phillips, etc., and continues its developing through last decades with the contribution of mathematicians as Grothendieck, Dinculeanu, Diestel and Uhl, Kwapień, Maurey, Pisier, Bourgain, Talagrand, and many more.

We focus our attention here on $L_p(\mu, X)$ and $C(K, X)$ spaces (for notations, see below "Some Notations and Conventions"), because we believe they are the most important and representative examples of vector-valued function Banach spaces. They are the points of reference in any study on the subject. We are interested in a problem (or collection of problems) which has deserved a lot of attention in last years. To be precise we are interested in the following:

Problem 1: *When does $L_p(\mu, X)$ contain a copy of c_0, ℓ_1 or ℓ_∞? What about complemented copies? Same questions about $C(K, X)$.*

The aim of this monograph is to provide a detailed exposition of the answers to these questions, giving an unified and self-contained treatment. It continues in some way a small part of the big work intiated in Diestel & Uhl's "Vector Measures" [39], which in turn was a continuation of some chapters of Dunford & Schwartz's "Linear Operators" (Part I) [51].

Of course, the questions in Problem 1 are just the classical ones concerning the study of the subspaces of a given Banach space. So we are led to the Bessaga-Pełczyński's and Rosenthal's old theorems characterizing the Banach spaces which have copies of c_0 or ℓ_1 (see [35] or [93]). All these theorems give a very good insight in the internal structure of the Banach spaces involved. This is why we believe that Problem 1 is not only interesting by itself, but has also contributed very much to a better understanding of the structure of $L_p(\mu, X)$ and $C(K, X)$ spaces. In particular, it has made clear that their structure is much more than a simple addition of the structures of X and $L_p(\mu)$ or $C(K)$. The problem has also motivated the development of useful techniques, which are the fruit of the work of many authors during the last twenty-five years.

Let us add that we have now an almost complete solution to Problem 1.

It is interesting to have in mind a standard way of thinking in vector-valued function spaces. Since X and $L_p(\mu)$ are (isometrically isomorphic to) complemented subspaces of $L_p(\mu, X)$, it is clear that if either X or $L_p(\mu)$ contains a copy of a Banach space Y, then so does $L_p(\mu, X)$. Of course, the same can be said about *complemented* copies. So, very often our main problem is to know whether the converse is true or not. The analogous situation also happens for $C(K, X)$. This is why the first problem we usually have to face is the following:

Problem 2: *Assume F is any of the spaces: c_0, ℓ_1 or ℓ_∞. Determine which of the following implications are true*

$$L_p(\mu, X) \supset F \overset{?}{\Longrightarrow} L_p(\mu) \supset F \ \text{or} \ X \supset F$$

$$L_p(\mu, X) \underset{(c)}{\supset} F \overset{?}{\Longrightarrow} L_p(\mu) \underset{(c)}{\supset} F \ \text{or} \ X \underset{(c)}{\supset} F$$

Consider the analogous questions for $C(K, X)$, that is,

$$C(K, X) \supset F \overset{?}{\Longrightarrow} C(K) \supset F \ \text{or} \ X \supset F$$

$$C(K, X) \underset{(c)}{\supset} F \overset{?}{\Longrightarrow} C(K) \underset{(c)}{\supset} F \ \text{or} \ X \underset{(c)}{\supset} F$$

When one of the implications in Problem 2 is true, we obtain a complete and natural solution to Problem 1 of the following kind:

$$L_p(\mu, X) \supset \ell_1 \Longleftrightarrow L_p(\mu) \supset \ell_1 \ \text{or} \ X \supset \ell_1$$

We will see that in many cases, but not always, the answers are analogous to the preceding one.

Let us give a look to the content of the monograph.

In Chapter 1, Preliminaries, we give some fundamental results which we will need. We have done an effort in the careful selection of these results. In the first three sections we recall the main characterizations of Banach spaces containing copies (or *complemented* copies) of c_0, ℓ_1 or ℓ_∞. In the remaining four sections we include different basic facts about $L_p(\mu, X)$ and $C(K, X)$ spaces. In particular, in Section 1.5 we study in detail the representation of the dual of $L_p(\mu, X)$ spaces provided by lifting theory. Of course, the main references here are Dinuleanu's "Vector Measures" [40] and A. and C. Ionescu Tulcea's "Topics in the Theory of Liftings" [75]. However, in these books some results on the dual of $L_p(\mu, X)$ spaces are quite dispersed. This is why we have preferred to provided quite a complete and unified treatment of the subject.

The Second Chapter is devoted to Kwapień's and Pisier's Theorems characterizing when $L_p(\mu, X)$ contains copies of c_0 and ℓ_1, respectively. These theorems were proved in the mid seventies and they were the first solutions to our problems. They are deep and difficult, and for this reason they take an important part of our monograph. Concerning Kwapień's Theorem, we should remark the important contribution of Hoffmann-Jørgensen with his preliminary work. We will explain this better in the Notes and Remarks of the Chapter. To prove the theorem we do not follow Kwapień's approach but Bourgain's. This is very important for us because we will need Bourgain's results in subsequent Chapters.

Chapter Three is devoted to $C(K, X)$ spaces. Theorems of Saab and Saab, Cembranos-Freniche and Drewnowski are studied. They were proved in the eighties and are concerned with complemented copies of ℓ_1, complemented copies of c_0 and copies of ℓ_∞, respectively. The easiest of them is Cembranos-Freniche's, which could be viewed as an exercise on Josefson-Nissenzweig theorem. However, it was quite surprising and had a particular influence when it was proved, because it provided the first negative answer to Problem 2 above. Notice that while Kwapień's and Pisier's Theorems were very difficult, they gave a natural answer.

Chapter Four is devoted to the problems on $L_p(\mu, X)$ spaces ($1 \leq p < +\infty$) not solved by Kwapień's and Pisier's Theorems: complemented copies of c_0, complemented copies of ℓ_1 and copies of ℓ_∞. We give here contributions of Bombal, Emmanuelle and Mendoza, obtained between 1988 and 1992. A curious fact to be mentioned here is that we find a clear difference between the behavior of purely atomic and *non* purely atomic measure spaces, in contrast with the results of the preceding Chapters. This happens when we consider *complemented* copies of c_0 in $L_p(\mu, X)$, and it will happen again in the next Chapter when dealing with $L_\infty(\mu, X)$.

Chapter Five is devoted to $L_\infty(\mu, X)$. We wish to characterize when this space contains complemented copies of c_0 or ℓ_1. This is shown by the results of Leung and Räbiger, Díaz and Kalton. They were obtained from 1990 until very recently (see the "Notes and Remarks" of the Chapter). In the ℓ_1 case, we find a connection between local theory and the problems we are studying. We believe that this is the last important finding in our subject.

Next, we have included a table summarizing the results of the monograph, that is, the solutions to Problem 1. We see that we have a complete solution to our problem with the only exception of the $L_\infty(\mu, X)$ case, and even in this case, for finite (or σ-finite) measures the problem is completely solved.

Finally, we have devoted a short Chapter to comment some open and related problems of the theory. It is clear that in our original Problem 1, we can change the spaces $L_p(\mu, X)$ and $C(K, X)$ for some other more or less similar spaces, like Köthe-Bochner spaces or tensor products, to mention two simple examples. On the other hand, we can also change the candidate subspaces c_0, ℓ_1 or ℓ_∞ for some other else, like ℓ_p $(1 < p < +\infty)$ or $L_1([0,1])$. There are also connections between the problems we have been considering and some other problems, related to classical properties of Banach spaces. In this final chapter, we comment some of these aspects.

This work is intended for those mathematicians interested in Banach Space Theory, Functional Analysis, and in general Abstract Analysis. The subject is crossroad of many branches in Banach space theory, and so one has occasion of applying many fundamental and classical results. For this reason, we think that the techniques developed here are interesting for anyone working on Banach spaces and absolutely fundamental for those interested in the study of vector-valued function spaces and related fields.

It is written at a graduate student level. In fact we have followed the first draft of this work to teach our 30 hours Seminar during the Course 96-97, and we have taken a great advantage of that experience. We think that it might be a first step to begin a research in the field. We assume the reader knows the basics on Banach space theory, and we mean for this to be familiar, for instance, with most parts of J. Diestel's "Sequences and Series in Banach Spaces " [35].

Of course, many of the proofs given here are different to the original ones. And we believe that some of them are much easier. In fact, we have done an effort trying to give the best points of view and the best proofs to understand the theory. We would say that in some cases we give surprisingly easy proofs and one could think that we are actually proving almost trivial results. However, if one reads the original papers one does not have the same impression. We would say that this is so because of two main reasons.

On one hand, as we have already mentioned, we have chosen carefully the preliminary results we need, and even the precise versions which are the most suitable for our purposes. Although these results are "well known", most of them were quite dispersed and very often they could not be found easily in the literature.

On the other hand, it is normal that after several years we can see a theory in a much clearer way, that is, we understand better the results. When they were discovered, neither their relationship with other results nor the right tools to prove them were so clear. In our case for instance, Saab & Saab's Theorem 3.1.4 on complemented copies of ℓ_1 might seem now to be very easy. However, if we reflect a little bit about the main ingredients in its proof: lifting theory and Bourgain's Theorem 2.1.1, one realizes immediately that in the early eighties this theorem of Bourgain was not very well known, and it was not at all usual to apply lifting theory to get a theorem on $C(K, X)$ spaces.

Bourgain's name has just appeared. We would like to emphasize that our monograph contains a detailed exposition of some fundamental results of Bourgain which were not very accessible until now, at least for a beginner in the subject. His ideas are all over this monograph and are very important in our study.

We must add that this is not just a compillation of known results. We have tried to put some important ideas into an accessible expository form and we have also included: (a) some results which appear here for the first time, and (b) some results which we believe have been quite unnoticed and are difficult to find in the literature. We give now a list of the most important of these results. The first three are in case (b), and the other two in case (a).

1. Proposition 1.6.3 on measurability. The result is quite an easy application of fundamental results about the Souslin operation, but we have not seen such an application in this context in the literature.

2. Lemma 2.1.3. It is a result due to Rosenthal which we have called Kadec-Pełczyński-Rosenthal's Lemma because it is a refinement of known ideas of the classical Kadec-Pełczyński paper [77]. We think that at the moment it is the best "subsequence splitting lemma" in $L_1(\mu)$.

3. Theorem 2.1.4. It gives a large class of subspaces of $L_p(\mu, X)$ in which the L_p-norm and the L_1-norm are equivalent. We obtain the result as a quite easy consequence of the just mentioned Kadec-Pełczyński-Rosenthal's Lemma 2.1.3.

4. Díaz-Kalton's Theorem 5.2.3 characterizing when $L_\infty(\mu, X)$ contains a complemented copy of ℓ_1. When we write these lines, neither Díaz's contribution nor Kalton's have been published yet. However, they kindly have allowed us to use their ideas and preprints to give a complete account of their results.

5. Theorem characterizing when $L_\infty(\mu, X)$ contains a complemented copy of $L_1([0, 1])$. It may be found in the Notes and Remarks of Chapter 5. It is

actually an immediate consequence of Díaz-Kalton's Theorem 5.2.3 and Hagler-Stegall's theorem [67] characterizing those duals which contain a complemented copy of $C([0,1])^*$.

ACKNOWLEDGEMENTS

First of all we wish to express our gratitude to F. Bombal, our teacher, because he guided our first steps in the research and introduced us in this subject. We also have to thank B. Rodríguez-Salinas, because he also had a great influence in our formation. J. Diestel's books, conferences, seminars and friendly conversations have been very important for us, too.

We may say that the origin of this monograph goes back to November of 1990. Then, K. Sundaresan invited the second named author to deliver a talk at Cleveland State University. The subject of that talk coincided exactly with the one of this monograph and K. Sundaresan encouraged him to continue to work on this field. A few years later, the same author was invited by K. Jarosz to deliver a talk at the Second Conference on Function Spaces held at Edwarsville in the Spring of 1994. This was a new motivation to continue and bring to date the work which had been exposed at Cleveland. As a result he wrote the survey [98], which has become a very condensed preliminary version of this monograph. For these reasons, we are indebted with K. Sundaresan and K. Jarosz for their encouragement and contribution to the birth of this work.

We are also indebted with J.M.F. Castillo who, may be unaware, encouraged us to write this monograph.

We have already mentioned above that N. J. Kalton and S. Díaz have allowed us to include here some important results of them which will appear here for the first time. We have also had several interesting conversations with them on different aspects of this work. We are indeed very grateful to them.

We are also grateful to many colleagues which have gently provided us some preprints and with whom we have had interesting conversations. Among others, we have to mention: L. Drewnowski, Z. Hu and B.-L. Lin.

J. Bourgain and H. Rosenthal kindly told us about the history of the key lemma 2.1.3, since they knew we were wrong about its origin. H. Rosenthal gave us very interesting advices, too. They have our gratitude.

We have to thank I. Villanueva, student of our seminar of 96-97, for several valuable suggestions and comments.

It is a pleasure to recall also our colleagues and friends of our Department who have helped and encouraged us very often.

J. Esquinas answered very patiently many questions about TeX we posed him. Also Springer-Verlag staff, and particularly, F. Holzwarth, gave us all the help we asked them.

Finally, during the preparation of the manuscript we have asked D. H. Fremlin about different aspects of our work. He has always been extremely kind and his help can not be overestimated. We are very grateful to him.

Of course, we are the only responsible of any possible mistakes in this work.

This work has been partially supported by D.G.I.C.Y.T. grant PB94-0243

Madrid, September 1997

Some Notations and Conventions

In general, we will use standard notation as in [35], [39] or [93].

We will work with real or complex Banach spaces.

Let X, Y be Banach spaces. As usual, we say that X has a (complemented) copy of Y if X has a (complemented) subspace which is isomorphic to Y. We will denote

$$X \supset Y \quad \text{and} \quad \underset{(c)}{X \supset Y}$$

Let X be a Banach space, let (Ω, Σ, μ) be a positive measure space , and let $1 \leq p \leq \infty$. We denote by $L_p(\mu, X)$ the Banach space of all X-valued p-Bochner μ-integrable (μ-essentially bounded, when $p = \infty$) functions with its usual norm. That is, the vectors of $L_p(\mu, X)$ are (equivalence classes of) μ-measurable functions f such that

$$\| f \|_p = \left(\int_\Omega \| f(\omega) \| \, d\mu(\omega) \right)^{\frac{1}{p}} < +\infty$$

in the case $1 \leq p < +\infty$, and

$$\| f \|_\infty = \text{ess sup} \{ \| f(\omega) \| : \omega \in \Omega \} < +\infty$$

in the case $p = +\infty$. In case X is the scalar field we simply denote $L_p(\mu)$.

If K is a compact Hausdorff space, we denote by $C(K, X)$ the Banach space of all continuous X-valued functions defined on K, endowed with the supremum norm. In case X is the scalar field we simply denote $C(K)$.

In order to avoid trivial situations, *we will always suppose that X, $C(K)$ and $L_p(\mu)$ are infinite dimensional.* Notice that in the case of $C(K)$ this simply means that K is infinite, and in the case of $L_p(\mu)$, that the corresponding measure space (Ω, Σ, μ) is not trivial, i.e. it has infinitely many disjoint measurable sets of finite positive measure.

For a vector measure $m : \Sigma \to X$, $\|m\|$ and $\tilde{m}$ denote the variation and semivariation of m respectively. The symbol $cabv(\Sigma, X)$ stands for the Banach space of all X-valued countably additive measures with bounded variation defined on Σ, endowed with the variation norm. If X is the scalar field this space is simply denoted by $cabv(\Sigma)$.

Recall that if (Ω, Σ, μ) is a measure space and we consider its completion $(\Omega, \bar{\Sigma}, \bar{\mu})$, then the corresponding $L_p(\mu, X)$ and $L_p(\bar{\mu}, X)$ spaces coincide. For this reason *we will always assume our measure spaces are complete.*

1. Preliminaries

Since we wish to study when the spaces $L_p(\mu, X)$ or $C(K, X)$ contain copies or complemented copies of c_0, ℓ_1 or ℓ_∞, we will devote the first part of the chapter to recall the main characterizations of when a general Banach space enjoys any of these properties. In the second part of the chapter we will give some general facts about the spaces $L_p(\mu, X)$ and $C(K, X)$.

1.1 Banach spaces containing c_0

Bessaga and Pełczyński's classical results [4] provide the main characterizations of Banach spaces containing c_0 we will need in our work (see [35, Chapter V]).

The basic sequences equivalent to the canonical basis of c_0 will be simply called c_0-*sequences*. Later on we will also use the expressions "ℓ_1-*sequence*" and in general "ℓ_p-*sequence*" in the same obvious sense. We will say that a sequence in X is *complemented* if its closed linear span is a complemented subspace of X. As usual, when we say that (x_n) is *seminormalized* we mean that $0 < inf \ \|x_n\| \le sup \ \|x_n\| < +\infty$.

Let us remember that a series $\sum x_n$ in X is called *weakly unconditionally Cauchy* (or *weakly unconditionally convergent*), or in short w.u.C., if $\sum x^*(x_n)$ is absolutely convergent for each $x^* \in X^*$ (see [35, Theorem 6, Chapter V] for the main properties of w.u.C. series). The typical example of a non trivial w.u.C. series is the canonical c_0-basis, moreover we have (see [35, pages 42-45]):

Theorem 1.1.1 (Bessaga-Pełczyński). *Let* (x_n) *be a seminormalized sequence and suppose that the series* $\sum x_n$ *is w.u.C., then* (x_n) *has a* c_0-*subsequence. Moreover, if* (x_n) *is a basic sequence, then it is a* c_0-*sequence.*

We also can give a characterization of the Banach spaces containing *complemented* copies of c_0. We need first the following simple result

Proposition 1.1.1 (Chapter VII, Exercise 4 of [35]). *The bounded linear operators from* X *into* c_0 *correspond precisely to the weak*-null sequences in* X^*, *where each weak*-null sequence* (x_n^*) *in* X^* *has associated the operator* $x \mapsto (x_n^*(x))$.

With the preceding proposition in mind the following result is trivial (as usual, $\delta_{nm} = 0$ if $n \neq m$ and $\delta_{nn} = 1$).

Proposition 1.1.2. *Let (x_n) be a c_0-sequence in X, then (x_n) is complemented if and only if there is a weak*-null sequence (x_n^*) in X^* such that*

$$x_n^*(x_m) = \delta_{nm}.$$

But the main criterion we will use to find complemented c_0-sequences will be the following (versions of it may be found in [52] or [122]):

Theorem 1.1.2. *Let $\sum x_n$ be a w.u.C. series in X, then (x_n) has a complemented c_0-subsequence if and only if there exists a weak*-null sequence (x_n^*) in X^* such that*

$$x_n^*(x_n) \nrightarrow 0.$$

Proof. Let us show the non trivial implication. Let $\sum x_n$ be a w.u.C. series in X, and let (x_n^*) be a weak*-null sequence in X^* such that

$$x_n^*(x_n) \nrightarrow 0.$$

Then, without loss of generality, we may assume that there exists $\delta > 0$ such that

$$| \, x_n^*(x_n) \, | > \delta$$

for all $n \in \mathbb{N}$. Let T be the bounded linear operator associated to (x_n^*) as in Proposition 1.1.1. Notice that $\sum x_n$ and $\sum T(x_n)$ satisfy the assumptions of Theorem 1.1.1, and then, taking subsequences if necessary, we may assume that (x_n) and $(T(x_n))$ are both c_0-sequences. Of course this says that $T|_F$ is an isomorphism, where F denotes the closed linear span of the x_n's. Observe now that Sobczyk's theorem implies that the closed linear span of the $T(x_n)$'s is complemented in c_0. Let P be a projection from c_0 onto this subspace. It is clear that $(T|_F)^{-1} \circ P \circ T$ is a projection from X onto F.

1.2 Banach spaces containing ℓ_1

Of course the main characterization of Banach spaces containing ℓ_1 is the classical Rosenthal's theorem (see Chapter XI of [35]). To give a criterion to find *complemented* ℓ_1-sequences we will use again w.u.C. series. The following result is quite easy.

Proposition 1.2.1 (Exercise 3 of Chapter VII of [35]). *The bounded linear operators from a Banach space X into ℓ_1 correspond precisely to the w.u.C. series in X^*, where each w.u.C. series $\sum x_n^*$ in X^* has associated the operator $x \mapsto (x_n^*(x))$.*

From the preceding result we get immediately the following:

Proposition 1.2.2. *Let (x_n) be an ℓ_1-sequence in X, then (x_n) is complemented if and only if there exists a w.u.C. series $\sum x_n^*$ in X^* such that*

$$x_n^*(x_m) = \delta_{nm}.$$

But to give a really useful criterion to find *complemented* ℓ_1-sequences we need another fundamental result of Rosenthal [110] which has many different applications: the disjointification lemma. Its proof may be found in [35, page 82].

Lemma 1.2.1 (Rosenthal's Disjointification lemma). *Let (μ_n) be a bounded sequence in $cabv(\Sigma)$. Then given $\epsilon > 0$ and a sequence (A_n) of pairwise disjoint members of Σ there exists and increasing sequence (k_n) of positive integers for which*

$$\|\mu_{k_n}\| \left(\bigcup_{j \neq n} A_{k_j} \right) < \epsilon$$

for all n.

We can now give the announced criterion. It is very similar to the one we have given for c_0-sequences (Theorem 1.1.2) in the preceding section.

Theorem 1.2.1 (Rosenthal [110, 111]). *Let (x_n) be a bounded sequence in X. Then (x_n) has a complemented ℓ_1-subsequence if and only if there exists a weakly unconditionally Cauchy series $\sum x_n^*$ in X^* such that*

$$x_n^*(x_n) \not\to 0.$$

Proof. The condition is of course necessary. For the converse take (x_n) and (x_n^*) as in the statement. We may suppose, without loss of generality, that both sequences are in the corresponding unit balls. Taking subsequences and multiplying the x_n^*'s by adequate scalars if necessary, we may also assume that there exists $\delta > 0$ such that

$$x_n^*(x_n) \geq \delta$$

for all n. Let $\mathcal{P}(\mathbb{N})$ be the σ-algebra of all subsets of $\mathbb{N}$. For each $m \in \mathbb{N}$ consider the countably additive scalar measure μ_m defined on $\mathcal{P}(\mathbb{N})$ by

$$\mu_m : \mathcal{P}(\mathbb{N}) \quad \to \quad \mathbb{K}$$
$$A \quad \to \quad \sum_{n \in A} x_n^*(x_m)$$

(this is just the usual way in which an element of ℓ_1 may be viewed as a member of $cabv(\mathcal{P}(\mathbb{N}))$). Since $\sum x_n^*$ is a w.u.C. series in X^* and (x_n) is bounded, using Proposition 1.2.1 it is clear that (μ_m) is bounded in $cabv(\mathcal{P}(\mathbb{N}))$. Hence we deduce from the preceding lemma that there are subsequences of (x_n) and (x_n^*), which we continue to denote again in the same way, such that

$$\|\mu_m\| \left(\bigcup_{n \neq m} \{n\} \right) = \sum_{n \neq m} |x_n^*(x_m)| < \delta/2$$

for all m. Let $R : X \to \ell_1$ be the bounded linear operator associated to $\sum x_n^*$ as in Proposition 1.2.1. Put $z_m = R(x_m) = (x_n^*(x_m))_n$. It is straightforward that the bounded linear operator

$$\begin{aligned} S : \ell_1 &\longrightarrow \ell_1 \\ (t_m) &\longrightarrow \sum t_m z_m \end{aligned}$$

satisfies

$$\| I - S \| < 1 - \frac{\delta}{2},$$

where I is the identity operator in ℓ_1. Hence S is an isomorphism and we can deduce that $(z_m) = (R(x_m))$ is an ℓ_1-sequence. Therefore (x_m) is an ℓ_1-sequence, too. Finally, if we define $T : \ell_1 \to X$ by $T((t_m)) = \sum t_m x_m$, it is immediate that $T \circ S^{-1} \circ R$ is a projection from X onto the closed linear span of the x_n's.

1.3 Banach spaces containing ℓ_∞

The main criteria to determine if a given Banach space contains ℓ_∞ are also due to Rosenthal [110].

We denote by (e_n) the unit vector sequence of ℓ_∞.

Theorem 1.3.1. *The following are equivalent:*

(a) $X \supset \ell_\infty$.

(b) *There exists a bounded linear operator $T : \ell_\infty \to X$ such that*

$$\| T(e_n) \| \not\to 0.$$

(c) *There exists a bounded linear operator $T : \ell_\infty \to X$ which is not weakly compact.*

Observe that (c) implies (a) is the only non easy implication in this theorem (and it is just VI.1.3. of [39]). That (a) implies (b) is trivial, and to understand why (b) implies (c) it is enough to recall that every weakly compact operator is unconditionally converging (that is, it transforms w.u.C. series into unconditionally convergent ones [35, Chapter V, Exercise 8]). Since $\sum e_n$ is a w.u.C. series in ℓ_∞, it is clear that the operator T of condition (b) is not unconditionally converging, and so it can not be weakly compact either.

Remark 1.3.1. If A is a set, we denote by $[A]$ the set of all infinite subsets of A. For $M \in [\mathbb{N}]$, $\ell_\infty(M)$ is defined to be the subspace of ℓ_∞ of all sequences $(\zeta_n) \in \ell_\infty$ with $\zeta_n = 0$ for $n \notin M$. Of course it is isometrically isomorphic to ℓ_∞. It is worth to mention that if T satisfies (b) of the preceding theorem then there exists $M \in [\mathbb{N}]$ such that $T|_{\ell_\infty(M)}$ is an isomorphic embedding (see Remark 1 of Proposition 1.2 of [110]).

A reformulation of (c) implies (a) in the preceding theorem is the following

Corollary 1.3.1. *If X has no copy of ℓ_∞ then every bounded linear operator $T : \ell_\infty \to X$ is weakly compact.*

And we get from this corollary the following classical result:

Corollary 1.3.2 (Phillips, 1940). *ℓ_∞ has no complemented copies of c_0.*

At this point it is interesting to know how are the operators from ℓ_∞ in X when X *does not* have copies of ℓ_∞. This is our purpose now.

Remark 1.3.2. We have mentioned above that weakly compact operators are unconditionally converging. Therefore, if an operator $T : \ell_\infty \to X$ is weakly compact then $\sum T(e_n)$ is unconditionally convergent, and so $\sum \lambda_n T(e_n)$ converges for all $(\lambda_n) \in \ell_\infty$. Of course, this does not mean that $T((\lambda_n))$ and $\sum_{n=1}^\infty \lambda_n T(e_n)$ coincide. However, we will see in the next proposition that this is "almost" true. In its proof we will need to take a family $\{N_t\}_{t \in [0,1]}$ of infinite subsets of $\mathbb{N}$ such that for each pair of different real numbers $s, t \in [0,1]$ the set $N_s \cap N_t$ is finite. How to show that such a family exists ? : Of course it suffices to show this for $\mathbb{Q}$ instead of $\mathbb{N}$. Well, for each $t \in [0,1]$ take N_t the set of values of a sequence of different rational numbers converging to t. It is clear that the family $\{N_t\}_{t \in [0,1]}$ enjoys the required property.

Proposition 1.3.1 (Drewnowski [46]). *Let $T : \ell_\infty \to X$ be a weakly compact operator, then there exists an infinite subset M of $\mathbb{N}$ such that*

$$T((\zeta_n)) = \sum_{n=1}^\infty \zeta_n T(e_n) \tag{1.1}$$

for all $(\zeta_n) \in \ell_\infty(M)$.

Proof. If $T : \ell_\infty \to X$ is weakly compact we have just seen that $\sum T(e_n)$ is unconditionally convergent. Thus the operator $S : \ell_\infty \to X$ given by $S((\xi_n)) = \sum \xi_n T(e_n)$ for $(\xi_n) \in \ell_\infty$, is well defined. Let m and m_c be the representing measures of T and S respectively, that is, the X-valued measures defined on $\mathcal{P}(\mathbb{N})$ (the set of all subsets of $\mathbb{N}$) by

$$m(A) = T(\chi_A) \quad \text{and} \quad m_c(A) = S(\chi_A) = \sum_{n \in A} T(e_n)$$

for each $A \in \mathcal{P}(\mathbb{N})$, where χ_A is the sequence (t_n) with $t_n = 1$ if $n \in A$ and $t_n = 0$ otherwise. Both measures are strongly additive: the first one because T is weakly compact [39, VI.1.1.], and the second one because it is even countably additive (it is the "countably additve part" of m). Therefore $m_0 = m - m_c$ is strongly additive, too. Remember that the strong additivity of m_0 is equivalent to the following property of its semivariation $\widetilde{m_0}$ [39, I.1.18.]: $\widetilde{m_0}(A_n) \to 0$ whenever (A_n) is a sequence of pairwise disjoint subsets of $\mathbb{N}$. By the argument given in the previous remark we know that there exists a family $\{N_t\}_{t \in [0,1]}$ of infinite subsets of $\mathbb{N}$ such that $N_t \cap N_s$ is finite if $t \neq s$. Since $\widetilde{m_0}$ is a non decreasing subadditive set function which vanishes on finite sets, it is immediate that $\widetilde{m_0}(N_1) = \widetilde{m_0}(N_2)$ if N_1 and N_2 differ only in a finite set. So, for each $\epsilon > 0$ the set

$$\{t \in [0,1] : \widetilde{m_0}(N_t) > \epsilon\}$$

is finite, and consequently $\{t \in [0,1] : \widetilde{m_0}(N_t) > 0\}$ is countable. Thus, there exists $s \in [0,1]$ such that $\widetilde{m_0}(N_s) = 0$. Of course this means that m_0 vanishes on all subsets of N_s, and taking $M = N_s$, it is clear that T and S coincide in $\ell_\infty(M)$ as required.

Corollary 1.3.3 (Drewnowski [46]). *Let (T_k) be a sequence of weakly compact operators from ℓ_∞ into X, then there exists an infinite subset M of $\mathbb{N}$ such that*

$$T_k((\zeta_n)) = \sum_{n=1}^{\infty} \zeta_n T_k(e_n)$$

for all $(\zeta_n) \in \ell_\infty(M)$ and all $k \in \mathbb{N}$.

Proof. We just have to use the preceding proposition and a diagonal argument. Apply the preceding proposition to T_1. We get $M_1 \in [\mathbb{N}]$ such that T_1 satisfies 1.1 in $\ell_\infty(M_1)$. Now, apply again the preceding proposition to T_2 (as defined in $\ell_\infty(M_1)$). We get $M_2 \in [M_1]$ such that T_2 satisfies 1.1 in $\ell_\infty(M_2)$. In this way we get a non increasing sequence (M_n) of infinite subsets of $\mathbb{N}$ such that each T_n satisfies 1.1 in $\ell_\infty(M_n)$. Let M be the subset of $\mathbb{N}$ consisting of the first natural number of M_1, the second one of M_2, and so on. It is clear that M satisfies the required property.

1.4 Natural subspaces of $L_p(\mu, X)$ and $C(K, X)$

Let us give our first glance to $L_p(\mu, X)$ and $C(K, X)$. Since we are interested in the study of subspaces of these spaces, let us begin recalling which are the obvious ones. Of course they are $L_p(\mu)$, X and ℓ_p in the case of $L_p(\mu, X)$, and $C(K)$, X and c_0 in the case of $C(K, X)$. We devote this short section to remember these elementary facts. They are very easy, but we believe that some reflection on them helps to understand better the spaces we are dealing with.

For $1 \leq p < +\infty$, take any vector x_0 in the unit sphere of X, then the map

$$
\begin{aligned}
L_p(\mu) &\longrightarrow L_p(\mu, X) \\
f &\longrightarrow f(.)x_0
\end{aligned}
$$

is an isometric embedding. Moreover, its range is norm one complemented in $L_p(\mu, X)$. To give a norm one projection it is enough to take any norm one functional x_0^* on X such that $<x_0, x_0^*> = 1$, and consider the map

$$
\begin{aligned}
L_p(\mu, X) &\longrightarrow L_p(\mu, X) \\
f &\longrightarrow <f(.), x_0^*> x_0
\end{aligned}
$$

Similarly, if f_0 is any function in the unit sphere of $L_p(\mu)$, then the map

$$
\begin{aligned}
X &\longrightarrow L_p(\mu, X) \\
x &\longrightarrow f_0(.)x
\end{aligned}
$$

is an isometric embedding and its range is norm one complemented, too. This time, to get a norm one projection, take a function g_0 in the unit sphere of $L_q(\mu)$ such that

$$
\int_\Omega f_0(t)g_0(t)\,d\mu(t) = 1,
$$

then it is enough to define

$$
\begin{aligned}
L_p(\mu, X) &\longrightarrow L_p(\mu, X) \\
f &\longrightarrow f_0(.)\int_\Omega g_0(t)f(t)\,d\mu(t)
\end{aligned}
$$

The situation for $L_\infty(\mu, X)$ and $C(K, X)$ is completely analogous. We can see in the same way that $L_\infty(\mu, X)$ contains many norm one complemented isometric copies of $L_\infty(\mu)$ and X, and $C(K, X)$ contains many norm one complemented isometric copies of $C(K)$ and X.

On the other hand, $L_p(\mu)$ does have many complemented copies of ℓ_p. If we take a sequence (A_n) of pairwise disjoint measurable sets with finite positive measure, it is straightforward to show that

$$
\begin{aligned}
L_p(\mu) &\longrightarrow L_p(\mu) \\
f &\longrightarrow \sum_{k=1}^\infty \frac{1}{\mu(A_k)^{\frac{1}{p}}}\left(\int_{A_k} f(t)\,d\mu(t)\right)\chi_{A_k}(.)
\end{aligned}
$$

is a norm one linear projection onto the closed linear span of the χ_{A_k}'s, a subspace isometrically isomorphic to ℓ_p. More generally,

Proposition 1.4.1. *For $1 \le p < +\infty$, if (f_k) is a seminormalized sequence in $L_p(\mu, X)$ of disjointly supported functions, then it is a complemented ℓ_p-sequence in $L_p(\mu, X)$. More precisely, the closed linear span of the f_k's is isometrically isomorphic to ℓ_p and it is norm one complemented.*

Proof. It is straightforward to show that the closed linear span of the f_k's is isometrically isomorphic to ℓ_p. On the other hand, if (τ_k) is a sequence of norm one functionals in $L_p(\mu, X)^*$ such that $< f_k, \tau_k >= \| f_k \|_p$, and we denote $A_k = supp(f_k) = \{ \omega \in \Omega : f_k(\omega) \ne 0 \}$, then one can easily show also that

$$
\begin{array}{ccc}
L_p(\mu, X) & \longrightarrow & L_p(\mu, X) \\
f & \longrightarrow & \displaystyle\sum_{k=1}^{\infty} < \chi_{A_k} f, \tau_k > \dfrac{1}{\| f_k \|_p} f_k
\end{array}
$$

is a norm one linear projection onto the closed linear span of the f_k's.

The situation for $L_\infty(\mu, X)$ and $L_\infty(\mu)$ is quite similar. They contain many isometric copies of ℓ_∞. In fact, if (f_k) is any normalized sequence in $L_\infty(\mu, X)$ of disjointly supported functions then the natural map

$$
\begin{array}{ccc}
\ell_\infty & \longrightarrow & L_\infty(\mu, X) \\
(t_k) & \longrightarrow & \displaystyle\sum_{k=1}^{\infty} t_k f_k(.)
\end{array}
$$

is an isometric embedding.

Finally, let us see why $C(K, X)$ (and $C(K)$) contains many isometric copies of c_0. Take any sequence (G_n) of pairwise disjoint non empty open subsets of K, and a sequence (t_n) of points in K such that $t_n \in G_n$ for all $n \in \mathbb{N}$. Thanks to Urysohn's theorem, for each $n \in \mathbb{N}$ there exists a continuous function f_n on K, taking values in $[0, 1]$, such that $f_n(t_n) = 1$ and $f_n(K \setminus G_n) = 0$. It is easy to show that the closed linear span on $C(K)$ of the f_n's is isometrically isomorphic to c_0. Moreover, if (x_n) any sequence of non null vectors in X, and we denote $g_n = f_n(.)x_n$, then the the closed linear span on $C(K, X)$ of the g_n's is isometrically isomorphic to c_0. In fact, we have:

Proposition 1.4.2. *If (g_n) is a sequence in $C(K, X)$ of non null functions such that the sets*

$$\{ t \in K : g_n(t) \ne 0 \}$$

are pairwise disjoint, then the closed linear span of the g_n's is a subspace isometrically isomorphic to c_0.

Proof. Of course we may assume the sequence (g_n) is normalized, then it is enough to notice that

$$
\begin{array}{ccc}
c_0 & \longrightarrow & C(K, X) \\
& & \infty \\
(\lambda_n) & \longrightarrow & \sum_{n-1} \lambda_n g_n
\end{array}
$$

is an isometric embedding.

Remark 1.4.1. The preceding proposition in general does not provide a *complemented* copy of c_0. In fact, we should remember that $C(K)$ very often *does not* have complemented copies of c_0 (we will come back over this in Chapter 3).

1.5 Two Radon-Nikodým Theorems. The dual of $L_p(\mu, X)$

We wish to present here in detail how to get a Radon-Nikodým theorem for X^*-valued measures using lifting theory, and how to apply it to obtain a representation of the dual of $L_p(\mu, X)$. Both results are quite old. The first one is contained in [40] and the second one in [40] and [75]. However, one has to combine several results of those monographs to get precisely the versions we give (which are, of course, the ones we need). The inclusion of proofs will also help to understand better the theory we are studying.

A few words to recall a well known situation (see [35, Chapter IV.1.]). Assume the measure μ is finite. Given $\Gamma \in L_p(\mu, X)^*$ one can define a vector measure $m : \Sigma \to X^*$ by

$$
m(A)(x) = \Gamma(\chi_A(.)x)
$$

for all $x \in X$ and all $A \in \Sigma$. This measure turns out to be countably additive, of bounded variation and absolutely continuous with respect to μ. If we have a density function for m, that is, a measurable function $\varphi : \Omega \to X^*$ so that

$$
m(A) = \int_A \varphi(\omega) d\mu(\omega)
$$

for all $A \in \Sigma$, then we have

$$
\Gamma(f) = \int_\Omega <f(\omega), \varphi(\omega)> d\mu(\omega) \tag{1.2}
$$

for all $f \in L_p(\mu, X)$. Thus, the functional Γ is represented by the function φ (which, in this case, belongs to $L_q(\mu, X^*)$, with $\frac{1}{p} + \frac{1}{q} = 1$).

Nevertheless, we know that in general there is no such a density function for all Γ's (this happens only if X^* has the Radon-Nikodým property).

Now, what happens if we look for a density function in a weaker sense? For instance, what happens if we look for a w^*-measurable function $\varphi : \Omega \to X^*$ such that

$$ m(A)(x) = \int_A <x, \varphi(\omega)> d\mu(\omega) $$

for all $x \in X$ and all $A \in \Sigma$? Recall that we say that a function $\varphi : \Omega \to X^*$ is w^*-measurable if the map $\omega \mapsto <x, \varphi(\omega)>$ is measurable for all $x \in X$. We will see, using lifting theory, that such a w^*-measurable function *always* exists. Of course, in general it is not measurable but it shares many of the good properties of the functions belonging to $L_q(\mu, X^*)$, and it represents the functional Γ in the sense of 1.2.

Let us start with the definition of lifting in $\mathcal{L}_\infty(\lambda)$ and the main result of this theory: the existence theorem. As usual, for a positive measure λ on Σ, $\mathcal{L}_p(\lambda)$ denotes the vector space of all scalar, λ-measurable p-integrable (λ-essentially bounded for $p = +\infty$) functions defined on Ω (we *do not* identify here a function with its equivalence class).

Definition 1.5.1. *Let λ be a complete and positive measure. A lifting in $\mathcal{L}_\infty(\lambda)$ is a linear map $\rho : \mathcal{L}_\infty(\lambda) \to \mathcal{L}_\infty(\lambda)$ such that:*

1. *For each $f \in \mathcal{L}_\infty(\lambda)$, $\rho(f)(\omega) = f(\omega)$ for λ-almost all $\omega \in \Omega$.*
2. *If $f, g \in \mathcal{L}_\infty(\lambda)$ and $f(\omega) = g(\omega)$ for λ-almost all $\omega \in \Omega$, then $\rho(f)(\omega) = \rho(g)(\omega)$ for all $\omega \in \Omega$.*
3. *If $f, g \in \mathcal{L}_\infty(\lambda)$ and $f(\omega) \leq g(\omega)$ for λ-almost all $\omega \in \Omega$, then $\rho(f)(\omega) \leq \rho(g)(\omega)$ for all $\omega \in \Omega$.*
4. *Let $\alpha \in \mathbb{K}$. If $f \in \mathcal{L}_\infty(\lambda)$ and $f(\omega) = \alpha$ for λ-almost all $\omega \in \Omega$, then $\rho(f)(\omega) = \alpha$ for all $\omega \in \Omega$.*
5. *For all $f, g \in \mathcal{L}_\infty(\lambda)$, $\rho(fg)(\omega) = \rho(f)(\omega)\rho(g)(\omega)$ for all $\omega \in \Omega$ (multiplicativity).*

The meaning of condition 3 is clear if we work with $\mathbb{R}$ as scalar field. In the complex case the meaning is the obvious one: if $f, g \in \mathcal{L}_\infty(\lambda)$ are real valued and $f(\omega) \leq g(\omega)$ for λ-almost all $\omega \in \Omega$, then $\rho(f)$ and $\rho(g)$ are real-valued and $\rho(f)(\omega) \leq \rho(g)(\omega)$ for all $\omega \in \Omega$.

If we reflect for a moment on the preceding definition we see that one might consider a lifting as a mapping from $L_\infty(\lambda)$ in $\mathcal{L}_\infty(\lambda)$. From this point of view, it would be a linear right inverse of the canonical quotient map $\pi : \mathcal{L}_\infty(\lambda) \to L_\infty(\lambda)$ with nice properties.

Theorem 1.5.1 (Existence Theorem [40, 75]). *Let λ be a complete, finite and positive measure, then there exists a lifting in $\mathcal{L}_\infty(\lambda)$.*

Let λ be a complete, finite and positive measure on Σ. We denote by $\mathcal{L}_{w^*}^\infty(\lambda, X^*)$ the vector space of all functions $\varphi : \Omega \to X^*$, which are λ-essentially bounded and w^*-measurable, and by $cabv_\lambda(\Sigma, X^*)$ the vector space of all measures $m \in cabv(\Sigma, X^*)$ for which there is $\alpha \geq 0$ such that

$$\|m(A)\| \leq \alpha\lambda(A)$$

for all $A \in \Sigma$. Notice that, unlike most of the spaces we are dealing with, $cabv_\lambda(\Sigma, X^*)$ is a *non* complete subspace of $cabv(\Sigma, X^*)$.

Our first step consists in applying the lifting existence theorem to get a density function for each measure belonging to $cabv_\lambda(\Sigma, X^*)$. Subsequently, a standard application of this will allow us to find a density function for *all* λ-continuous measures in $cabv(\Sigma, X^*)$.

Theorem 1.5.2 (Radon-Nikodým type I). *Let λ be a complete, finite and positive measure on Σ. Each lifting ρ in $\mathcal{L}_\infty(\lambda)$ induces a linear and injective map*

$$\hat{\rho} : cabv_\lambda(\Sigma, X^*) \to \mathcal{L}_{w^*}^\infty(\lambda, X^*)$$

such that for each measure $m \in cabv_\lambda(\Sigma, X^*)$ the function $\varphi = \hat{\rho}(m)$ satisfies the following two conditions:

1. For each $A \in \Sigma$ and all $x \in X$

$$m(A)(x) = \int_A <x, \varphi(\omega)> d\lambda(\omega)$$

2. The function $\omega \mapsto \|\varphi(\omega)\|$ is measurable, and for each $A \in \Sigma$

$$\|m\|(A) = \int_A \|\varphi(\omega)\| d\lambda(\omega)$$

Proof. For simplicity we will give the proof in the real case. The result carries over in a standard manner to the complex case.

Let ρ be a lifting in $\mathcal{L}_\infty(\lambda)$. Let $m \in cabv_\lambda(\Sigma, X^*)$ and $\alpha > 0$ be such that

$$\|m(A)\| \leq \alpha\lambda(A) \tag{1.3}$$

for all $A \in \Sigma$. For each $x \in X$ let us define $m_x : \Sigma \to \mathbb{R}$ by

$$m_x(A) = m(A)(x)$$

for all $A \in \Sigma$. Of course, m_x belongs to $cabv(\Sigma)$. By 1.3, m_x is λ-continuous and so, by the Radon-Nikodým theorem, there exists $f_x \in \mathcal{L}_\infty(\lambda)$ such that

$$m_x(A) = \int_A f_x(\omega) d\lambda(\omega)$$

for all $A \in \Sigma$. Now, we can define $\varphi : \Omega \to X^*$ by

$$<x, \varphi(\omega)> = \rho(f_x)(\omega) \tag{1.4}$$

for all $x \in X$ and all $\omega \in \Omega$.

We must show that φ satisfies all our requirements. Recall that two density functions of the same scalar λ-continuous measure differ only in a λ-null set. With this in mind, thanks to the linearity of ρ, we have that $\varphi(\omega)$ is linear for each $\omega \in \Omega$. Let us see that it is also continuous. If we apply the Radon-Nikodým theorem to the measure $\|m\|$ we get a function $g \in \mathcal{L}_\infty(\lambda)$, with $\|g\|_\infty \leq \alpha$, such that

$$\|m\|(A) = \int_A g(\omega) \, d\lambda(\omega)$$

for each $A \in \Sigma$. Then, for all $A \in \Sigma$ and all $x \in X$, we have

$$\int_A f_x(\omega) \, d\lambda(\omega) \;\; = \;\; m_x(A) = m(A)(x) \leq \|m(A)\| \, \|x\|$$

$$\leq \;\; \|m\|(A) \, \|x\| = \left(\int_A g(\omega) \, d\lambda(\omega) \right) \|x\|$$

Therefore, for each $x \in X$,

$$f_x(\omega) \leq g(\omega) \, \|x\| \leq \alpha \, \|x\|$$

for λ-almost all $\omega \in \Omega$, and so

$$<x, \varphi(\omega)> = \rho(f_x)(\omega) \leq \rho(g)(\omega) \, \|x\| \leq \alpha \, \|x\| \tag{1.5}$$

whenever $\omega \in \Omega$ and $x \in X$. Thus, $\varphi(\omega) \in X^*$ and $\|\varphi(\omega)\| \leq \alpha$ for all $\omega \in \Omega$. In other words, φ is well defined and bounded. Moreover, since φ is clearly w^*-measurable, we have that $\varphi \in \mathcal{L}_{w^*}^\infty(\lambda, X^*)$. By its own definition it is also clear that φ satisfies condition 1.

To see that φ satisfies condition 2 we will show that $\|\varphi(\omega)\| = g(\omega)$ for λ-almost all $\omega \in \Omega$. By 1.5 we have

$$\|\varphi(\omega)\| \leq \rho(g)(\omega)$$

for all $\omega \in \Omega$, and so,

$$\|\varphi(\omega)\| \leq g(\omega)$$

for λ-almost all $\omega \in \Omega$. Let g_0 be the supremum in $L_1(\lambda)$ (or $L_\infty(\lambda)$, see [51, IV.11.7.]) of the set of all functions of the form

$$\sum_i \chi_{A_i}(\cdot) \, | \, \rho(f_{x_i})(\cdot) \, |$$

with $\{A_i\}$ a finite Σ-partition of Ω and $\{x_i\} \subset B(X)$. This function $g_0 \in \mathcal{L}_\infty(\lambda)$ satisfies

$$g_0(\omega) \leq \|\varphi(\omega)\| \leq g(\omega) \qquad (1.6)$$

for λ-almost all $\omega \in \Omega$. Given $\epsilon > 0$ take a finite Σ-partition $\{A_i\}$ of Ω and $\{x_i\} \subset B(X)$ such that

$$\|m\|\,(\Omega) - \epsilon \leq \sum m(A_i)(x_i).$$

Then

$$
\begin{aligned}
\int_\Omega g(\omega)\, d\lambda(\omega) - \epsilon &= \|m\|\,(\Omega) - \epsilon \leq \sum_i m(A_i)(x_i) \\
&= \sum_i \int_{A_i} <x_i, \varphi(\omega)>\, d\lambda(\omega) \\
&= \int_\Omega \left(\sum_i \chi_{A_i}(\omega) <x_i, \varphi(\omega)> \right) d\lambda(\omega) \\
&= \int_\Omega \left(\sum_i \chi_{A_i}(\omega) \rho(f_{x_i})(\omega) \right) d\lambda(\omega) \\
&\leq \int_\Omega g_0(\omega)\, d\lambda(\omega).
\end{aligned}
$$

Since this is true for all $\epsilon > 0$, we have

$$\int_\Omega g(\omega)\, d\lambda(\omega) \leq \int_\Omega g_0(\omega)\, d\lambda(\omega).$$

Therefore, using 1.6, we get

$$g_0(\omega) = \|\varphi(\omega)\| = g(\omega)$$

for almost all $\omega \in \Omega$.

Finally, from the definition of $\hat\rho(m) = \varphi$ given in 1.4 and the properties of a lifting, it is clear that $\hat\rho$ is linear and injective. This completes the proof.

Remark 1.5.1. Given $m \in cabv_\lambda(\Sigma, X^*)$, the function $\varphi = \hat\rho(m)$ is actually bounded (not only λ-essentially bounded). In fact, it is clear in the proof that if

$$\|m(A)\| \leq \alpha\lambda(A)$$

for all $A \in \Sigma$, then

$$\|\varphi(\omega)\| \leq \alpha$$

for all $\omega \in \Omega$.

In the first part of the proof it is explained how $\hat\rho$ is defined given a lifting ρ in $\mathcal{L}_\infty(\lambda)$. Actually the precise way in which $\hat\rho$ is defined will not be important in our study.

Of course $\hat{\rho}$ in the preceding theorem is not onto, but it should be noticed that each $\varphi \in \mathcal{L}_{w^*}^\infty(\lambda, X^*)$ defines a measure $m \in cabv_\lambda(\Sigma, X^*)$ in the natural way:

$$m(A)(x) = \int_A <x, \varphi(\omega)> d\lambda(\omega)$$

for each $A \in \Sigma$ and each $x \in X$.

Theorem 1.5.3 (Radon-Nikodým type II). *Let λ be a complete, finite and positive measure on Σ, and let $m \in cabv(\Sigma, X^*)$ be an absolutely continuous measure respect to λ. Then there exists $\Psi : \Omega \to X^*$ w^*-measurable satisfying the following three conditions:*

1. *The function $\omega \mapsto \|\Psi(\omega)\|$ (which we simply denote $\|\Psi(.)\|$) is measurable and belongs to $\mathcal{L}_1(\lambda)$*
2. *For all $x \in X$ and all $A \in \Sigma$*

$$m(A)(x) = \int_A <x, \Psi(\omega)> d\lambda(\omega)$$

3. *For all $A \in \Sigma$*

$$\|m\|(A) = \int_A \|\Psi(\omega)\| d\lambda(\omega)$$

Proof. Let $m \in cabv(\Sigma, X^*)$ be absolutely continuous respect to λ. Since the positive measure $\|m\|$ is also absolutely continuous respect to λ, by the scalar Radon-Nikodým theorem, there exists $g \in \mathcal{L}_1(\lambda)$ such that

$$\|m\|(A) = \int_A g(\omega) d\lambda(\omega)$$

for all $A \in \Sigma$. For each $n \in \mathbb{N}$ put $B_n = \{\omega \in \Omega : n - 1 \le g(\omega) < n\}$. The sequence (B_n) is a Σ-partition of Ω. Consider for each n the measure m_n defined by

$$m_n(A) = m(A \cap B_n)$$

for all $A \in \Sigma$. We have

$$\|m_n(A)\| = \|m(A \cap B_n)\| \le \|m\|(A \cap B_n) = \int_{A \cap B_n} g(\omega) d\lambda(\omega)$$
$$\le n\lambda(A \cap B_n) \le n\lambda(A)$$

for each $A \in \Sigma$. Therefore, m_n belongs to $cabv_\lambda(\Sigma, X^*)$. Let us fix a lifting ρ in $\mathcal{L}_\infty(\lambda)$, and let us consider the map $\hat{\rho}$ associated to it given in the preceding theorem. We can now define

$$\Psi : \Omega \longrightarrow X^*$$
$$\omega \longrightarrow \sum_{n=1}^{\infty} \chi_{B_n}(\omega)\hat{\rho}(m_n)(\omega)$$

It is clear that Ψ is w^*-measurable and $\|\Psi(.)\|$ is measurable. We also get that for all $A \in \Sigma$

$$
\begin{aligned}
\int_A \|\Psi(\omega)\| \, d\lambda(\omega) &= \sum_{n=1}^{\infty} \int_{A \cap B_n} \|\Psi(\omega)\| \, d\lambda(\omega) \\
&= \sum_{n=1}^{\infty} \|m_n\|(A) = \sum_{n=1}^{\infty} \|m\|(A \cap B_n) \\
&= \|m\|(A).
\end{aligned}
$$

This means that $\|\Psi(.)\|$ belongs to $\mathcal{L}_1(\lambda)$, and condition 3 holds. Finally, it is clear that for each $x \in X$ and all $A \in \Sigma$

$$
\begin{aligned}
\int_A <x, \Psi(\omega)> \, d\lambda(\omega) &= \sum_{n=1}^{\infty} \int_{A \cap B_n} <x, \Psi(\omega)> \, d\lambda(\omega) \\
&= \sum_{n=1}^{\infty} m(A \cap B_n)(x) = m(A)(x).
\end{aligned}
$$

Remark 1.5.2. After Theorem 1.5.2, one could think that the preceding proof provides a *linear* map from the subspace of $cabv(\Sigma, X^*)$ of all absolutely continuous measures with respect to λ into the vector space of w^*-measurable functions from Ω into X^*. However this is not true. The point is that our definition of Ψ given in the proof *does* depend on the choice of the B_n's. This might be surprising at first, but notice that if the definition of Ψ did not depend on the choice of the B_n's, the technique used in the proof could be used to show the existence of a linear lifting in $\mathcal{L}_p(\lambda)$ (a map from $\mathcal{L}_p(\lambda)$ into $\mathcal{L}_p(\lambda)$ with the same properties of a lifting with the exception of the multiplicativity condition 5). But it is well known that there are no linear liftings in $\mathcal{L}_p(\lambda)$ for $1 \le p < \infty$ (see [40, Proposition 9 of §13] or [75, Chapter III, Section 4]).

From the preceding Radon-Nikodým theorem it follows in quite a standard way the following theorem giving a representation of the continuous linear functionals in $L_p(\mu, X)$.

Theorem 1.5.4 (The dual of $L_p(\mu, X)$ I). *Let (Ω, Σ, μ) be a finite measure space, let $1 \le p < \infty$ and let q be the conjugate of p (i.e., $\frac{1}{p} + \frac{1}{q} = 1$), then for each $\Gamma \in L_p(\mu, X)^*$ there exists a function $\Psi : \Omega \to X^*$ such that:*

1. Ψ is w^*-measurable.

2. *The function* $\omega \mapsto \|\Psi(\omega)\|$ *(which we simply denote by* $\|\Psi(.)\|$*) is measurable and belongs to* $\mathcal{L}_q(\mu)$.

3. $\Gamma(f) = \int_\Omega <f(\omega), \Psi(\omega)> d\mu(\omega)$ *for all* $f \in L_p(\mu, X)$, *and*

4. $\|\Gamma\| = \|\|\Psi(.)\|\|_q$.

Conversely, each w^*-*measurable function* $\Psi : \Omega \to X^*$ *for which there exists* $g \in \mathcal{L}_q(\mu)$ *such that*

$$\|\Psi(\omega)\| \leq g(\omega)$$

for μ-*almost all* $\omega \in \Omega$, *induces by 3 a continuous linear functional* Γ *on* $L_p(\mu, X)$, *whose norm is less or equal to* $\|g\|_q$.

Proof. It is convenient to prove first the second part of the theorem, which is straightforward. In other words, it is easy to show that if you take $\Psi : \Omega \to X^*$ w^*-measurable, for which there exists $g \in \mathcal{L}_q(\mu)$ such that

$$\|\Psi(\omega)\| \leq g(\omega)$$

for μ-almost all $\omega \in \Omega$, then for each $f \in L_p(\mu, X)$ the function

$$\omega \mapsto <f(\omega), \Psi(\omega)>$$

belongs to $\mathcal{L}_1(\mu)$ and

$$
\begin{aligned}
L_p(\mu, X) &\longrightarrow \mathbb{K} \\
f &\longrightarrow \int_\Omega <f(\omega), \Psi(\omega)> d\mu(\omega)
\end{aligned}
$$

defines a continuous linear functional on $L_p(\mu, X)$, whose norm is less or equal to $\|g\|_q$.

Let us see now the first part of the theorem. Given $\Gamma \in L_p(\mu, X)^*$, let us consider the measure $m : \Sigma \to X^*$ associated to Γ in the usual way (see p. 98-99 of [39]):

$$m(A)(x) = \Gamma(\chi_A(.)x) \tag{1.7}$$

for all $A \in \Sigma$ and all $x \in X$. It is easy to show that this measure turns out to be an element of $cabv(\Sigma, X^*)$. Moreover, it verifies

$$\|m\|(A) \leq \mu(A)^{\frac{1}{p}} \|\Gamma\|$$

for all $A \in \Sigma$. Therefore m is absolutely continuous respect to μ and we can apply the preceding theorem. We get $\Psi : \Omega \to X^*$ w^*-measurable with $\|\Psi(.)\|$ belonging to $\mathcal{L}_1(\mu)$ such that

$$m(A)(x) = \int_A <x, \Psi(\omega)> d\mu(\omega) \tag{1.8}$$

for all $x \in X$ and all $A \in \Sigma$, and

$$\|m\|(A) = \int_A \|\Psi(\omega)\| \, d\mu(\omega) \tag{1.9}$$

for all $A \in \Sigma$. Obviously 1.8 means that for each simple functions f

$$\Gamma(f) = \int_\Omega <f(\omega), \Psi(\omega)> \, d\mu(\omega). \tag{1.10}$$

If we show that $\| \|\Psi(.)\| \|_q \leq \|\Gamma\|$ then we will have that condition 2 is satisfied. But then we can apply the second part of the theorem to the function Ψ and, thanks to 1.10, we can assure that Γ and the continuous linear functional associated to Ψ coincide in the simple functions. Therefore they coincide in the whole $L_p(\mu, X)$. Of course, this proves that conditions 3 is satisfied and also the inequality $\|\Gamma\| \leq \| \|\Psi(.)\| \|_q$. So, it only remains to prove $\| \|\Psi(.)\| \|_q \leq \|\Gamma\|$, and to do this it suffices to see that the measurable function $\| \Psi(.) \|$ defines in the usual way a continuos linear functional on $L_p(\mu)$ whose norm is bounded by $\|\Gamma\|$.

Take a scalar Σ-simple function $h = \sum_{i=1}^n \alpha_i \chi_{A_i}$. Given $\epsilon > 0$ there are, for each $i \in \{1, \ldots, n\}$, Σ-finite partitions $(B_j^i)_j$ of A_i, and vectors $(x_j^i)_j$ in the unit ball of X such that

$$\|m\|(A_i) \leq \sum_j m(B_j^i)(x_j^i) + \frac{\epsilon}{n}$$

Then, according to 1.7 and 1.9, we have

$$
\begin{aligned}
\left| \int_\Omega h(\omega) \|\Psi(\omega)\| \, d\mu(\omega) \right| &= \left| \sum_{i=1}^n \int_{A_i} \alpha_i \|\Psi(\omega)\| \, d\mu(\omega) \right| \\
&\leq \sum_{i=1}^n |\alpha_i| \|m\|(A_i) \\
&\leq \sum_{i=1}^n |\alpha_i| \left(\sum_j m(B_j^i)(x_j^i) + \frac{\epsilon}{n} \right) \\
&= \left(\sum_{i=1}^n \sum_j |\alpha_i| \, m(B_j^i)(x_j^i) \right) + \epsilon \\
&= \Gamma \left(\sum_{i=1}^n \sum_j \chi_{B_j^i}(.) \, |\alpha_i| \, x_j^i \right) + \epsilon \\
&\leq \|\Gamma\| \left\| \sum_{i=1}^n \sum_j \chi_{B_j^i}(.) \, |\alpha_i| \, x_j^i \right\|_p + \epsilon \\
&= \|\Gamma\| \left(\sum_{i=1}^n \sum_j |\alpha_i|^p \| x_j^i \|^p \, \mu(B_j^i) \right)^{\frac{1}{p}} + \epsilon
\end{aligned}
$$

$$\leq \quad \|\Gamma\| \left(\sum_{i=1}^{n} |\alpha_i|^p \, \mu(A_i) \right)^{\frac{1}{p}} + \epsilon$$

$$= \quad \|\Gamma\| \, \|h\|_p + \epsilon.$$

Since this is true for all $\epsilon > 0$ and all simple functions h we conclude that $\|\Psi(.)\|$ belongs to $\mathcal{L}_q(\mu)$ and $\| \|\Psi(.)\| \|_q \leq \|\Gamma\|$. This completes the proof.

Remark 1.5.3. Notice that the canonical map considered in the preceding proof

$$\begin{array}{ccc} L_p(\mu; X)^* & \longrightarrow & cabv(\Sigma, X^*) \\ \Gamma & \longrightarrow & m \end{array}$$

where $m(A)(x) = \Gamma(\chi_A(.)x)$ for all $A \in \Sigma$ and all $x \in X$, is an injective continuous linear operator from $L_p(\mu, X)^*$ into $cabv(\Sigma, X^*)$.

Theorem 1.5.4 gives us a good representation of *each* continuous linear functional in $L_p(\mu, X)$ and it is the result we will use in our work. If we now wish a representation of the *Banach* space $L_p(\mu, X)^*$, that is, if we wish to represent its linear structure, this is very easy. It is done in the following way. Let us denote by $\mathcal{L}_{w^*}^q(\mu, X^*)$ the vector space of all w^*-measurable functions $\Psi : \Omega \to X^*$ for which there exists $g \in \mathcal{L}_q(\mu)$ such that

$$\|\Psi(\omega)\| \leq g(\omega)$$

for μ-almost $\omega \in \Omega$. Consider in $\mathcal{L}_{w^*}^q(\mu, X^*)$ the equivalence relation $\overset{*}{\sim}$, where given $\Psi_1, \Psi_2 \in \mathcal{L}_{w^*}^q(\mu, X^*)$, we say that $\Psi_1 \overset{*}{\sim} \Psi_2$ if for each $x \in X$ we have

$$<x, \Psi_1(\omega)> \, = \, <x, \Psi_2(\omega)>$$

for μ-almost all $\omega \in \Omega$. Let us denote by $L_{w^*}^q(\mu, X^*)$ the quotient space and by $[\Psi]$ the equivalence classes. It is easy to show that $L_{w^*}^q(\mu, X^*)$ with the operations defined in the natural way is a vector space (that is, $\overset{*}{\sim}$ is compatible with the linear structure of $\mathcal{L}_{w^*}^q(\mu, X^*)$). If we define

$$\| [\Psi] \|_q = \inf \|g\|_q,$$

where the infimum is taken over all $g \in \mathcal{L}_q(\mu)$ for which there is some $\Psi_0 \in [\Psi]$ such that $\|\Psi_0(\omega)\| \leq g(\omega)$ μ-almost everywhere, it is also straightforward to show that $\| . \|_q$ is a norm of Banach space in $L_{w^*}^q(\mu, X^*)$ and that the natural embedding from $L_q(\mu, X^*)$ into $L_{w^*}^q(\mu, X^*)$ is an isometric isomorphism. Finally, let us consider the map

$$\begin{array}{ccc} I : L_{w^*}^q(\mu, X^*) & \longrightarrow & L_p(\mu, X)^* \\ [\Psi] & \longrightarrow & I([\Psi]) \end{array}$$

given by

$$I([\Psi])(f) = \int_\Omega < f(\omega), \Psi_0(\omega) > d\mu(\omega) \qquad (1.11)$$

for each $f \in L_p(\mu, X)$, if Ψ_0 is any function in $[\Psi]$. Standard arguments show that I is a well defined bounded linear operator. Moreover, the preceding theorem says that

Theorem 1.5.5 (The dual of $L_p(\mu, X)$ II). *Let (Ω, Σ, μ) be a finite measure space, and let $1 \le p < \infty$, then the bounded linear operator I defined in 1.11 is an isometric isomorphism onto $L_p(\mu, X)^*$, and for each $[\Psi] \in L_{w^*}^q(\mu, X^*)$ there exists $\Psi_0 \in [\Psi]$ such that the function $\omega \mapsto \| \Psi_0(\omega) \|$ (which we simply denote by $\| \Psi_0(.) \|$) is measurable and belongs to $L_q(\mu)$, and $\| | [\Psi] | \|_q = \| \, \| \Psi_0(.) \| \, \|_q$.*

1.6 Some remarks on $L_p(\mu, X)$ spaces

In this section we will give a few fundamental facts about $L_p(\mu, X)$ spaces. Our main aims are the following: (a) reducing our problems to the case of finite measures, (b) studying a question about measurability, (c) discussing briefly the differences between $L_p(\mu, X)$ spaces depending on the measure μ is purely atomic or not, and (d) recalling the definition of uniform integrability.

Many of the problems considered in $L_p(\mu, X)$, when (Ω, Σ, μ) is an *arbitrary* measure space, are solved as soon as we have the solution in the *finite* measure case. In particular, this is true for the problems studied here. Let us see why.

For two positive measure spaces (Ω, Σ, μ) and $(\Omega_1, \Sigma_1, \mu_1)$ we will denote $(\Omega_1, \Sigma_1, \mu_1) \subset (\Omega, \Sigma, \mu)$ if $\Omega_1 \in \Sigma$, Σ_1 is a sub-σ-algebra of Σ of subsets of Ω_1, and $\mu_1 = \mu|_{\Sigma_1}$. Of course we will identify $L_p(\mu_1, X)$ with a subspace of $L_p(\mu, X)$, just extending by zero the functions.

Remember that a measure space (Ω, Σ, μ) is said to be *separable* if the corresponding Banach space $L_1(\mu)$ (or equivalently $L_p(\mu)$, $1 < p < \infty$) is separable. It is easy to prove that this happens if and only if there exists a sequence (A_k) of Σ-measurable sets such that Σ is the μ-completion of the σ-algebra generated by the A_k's.

Let us begin with the following simple result which can be found in [51].

Lemma 1.6.1 (Lemma III.8.5. of [51]). *Let $1 \le p < \infty$ and let F be a separable subspace of $L_p(\mu, X)$, then there exists a σ-finite and separable measure space $(\Omega_1, \Sigma_1, \mu_1) \subset (\Omega, \Sigma, \mu)$ such that F lies in $L_p(\mu_1, X)$.*

Of course, if the space F in the preceding lemma is complemented in $L_p(\mu, X)$, then it is complemented in $L_p(\mu_1, X)$, too (in fact, the converse is also true as we will see in Lemma 1.6.2). So the lemma provides a reduction to the σ-finite case. Let us go further, to reach the finite measure case. The next proposition is well known.

Proposition 1.6.1. *Let $(\Omega_1, \Sigma_1, \mu_1)$ be a σ-finite measure space, then there exists a probability measure μ_0 on (Ω_1, Σ_1) such that for $1 \leq p \leq +\infty$, $L_p(\mu_1, X)$ and $L_p(\mu_0, X)$ are isometrically isomorphic. Besides, the measures μ_1 and μ_0 have the same null sets, and consequently they have the same atoms. In particular, μ_1 is purely atomic if and only if μ_0 is.*

Proof. Let (A_n) be a sequence of pairwise disjoint measurable sets with finite and positive μ_1-measure such that $\Omega_1 = \cup_n A_n$. Let us define

$$\mu_0 : \Sigma_1 \longrightarrow [0, +\infty)$$
$$A \longrightarrow \sum_{n=1}^{\infty} \frac{1}{2^n \mu_1(A_n)} \mu_1(A \cap A_n)$$

It is clear that μ_0 is a probability measure with the same null sets as μ_1, and so the identity

$$L_\infty(\mu_1, X) \longrightarrow L_\infty(\mu_0, X)$$
$$f \longrightarrow f$$

is an isometric isomorphism onto. Moreover, for $1 \leq p < \infty$ it is immediate that the map

$$L_p(\mu_1, X) \longrightarrow L_p(\mu_0, X)$$
$$f \longrightarrow hf$$

where

$$h = \sum_{n=1}^{\infty} 2^{\frac{n}{p}} \mu_1(A_n)^{\frac{1}{p}} \chi_{A_n}$$

is an isometric isomorphism onto, too.

As an immediate consequence of the preceding results we have:

Theorem 1.6.1. *Let (Ω, Σ, μ) be an arbitrary positive measure space, let $1 \leq p < \infty$, let F be a separable Banach space, and assume that $L_p(\mu, X)$ contains a copy (respectively, a complemented copy) of F. Then there exists a separable, finite and positive measure space $(\Omega_0, \Sigma_0, \mu_0)$ such that $L_p(\mu_0, X)$ contains a copy (respectively, a complemented copy) of F. Besides, if the measure space (Ω, Σ, μ) is purely atomic, we may assume that $(\Omega_0, \Sigma_0, \mu_0)$ has the same property.*

Remark 1.6.1. In this monograph we are mainly interested in the spaces $F = c_0$ and $F = \ell_1$.

Remark 1.6.2. Neither the preceding theorem nor Lemma 1.6.1 can be extended to *non separable* subspaces F of $L_p(\mu, X)$. To give simple examples showing this it is enough to take any measure space (Ω, Σ, μ) providing an $L_p(\mu)$ non separable and non isomorphic to an $L_p(\mu_0)$ with finite measure μ_0. Then consider as X the scalar field and as F the whole space $L_p(\mu)$.

Although, as we have just remarked, the above results are not true for general non separable spaces, it is very interesting for us that *they are true for ℓ_∞*. Our aim now is to show this. Let us begin with an easy result on complementability.

Lemma 1.6.2. *For $1 \leq p \leq \infty$, if $(\Omega_1, \Sigma_1, \mu_1) \subset (\Omega, \Sigma, \mu)$ is σ-finite, then $L_p(\mu_1, X)$ is norm one complemented in $L_p(\mu, X)$.*

Proof. Let us denote $\Sigma(\Omega_1) = \{A \in \Sigma : A \subset \Omega_1\}$. Clearly, $\Sigma(\Omega_1)$ is a sub-σ-algebra of Σ which contains Σ_1. Let us denote by R the restriction operator from $L_p(\mu, X) = L_p(\Omega, \Sigma, \mu; X)$ into $L_p(\Omega_1, \Sigma(\Omega_1), \mu|_{\Sigma(\Omega_1)}; X)$. Since $(\Omega_1, \Sigma(\Omega_1), \mu|_{\Sigma(\Omega_1)})$ is σ-finite we have a norm one conditional expectation operator E from $L_p(\Omega_1, \Sigma(\Omega_1), \mu|_{\Sigma(\Omega_1)}; X)$ into $L_p(\Omega_1, \Sigma_1, \mu_1; X) = L_p(\mu_1, X)$ (see V.1. of [39]). It is plain that $E \circ R$ is a norm one projection from $L_p(\mu, X)$ onto $L_p(\mu_1, X)$.

Proposition 1.6.2. *Let $1 \leq p < \infty$, and assume that $L_p(\mu, X)$ contains a copy of ℓ_∞. Then there exists a σ-finite and separable measure space $(\Omega_1, \Sigma_1, \mu_1) \subset (\Omega, \Sigma, \mu)$ such that $L_p(\mu_1, X)$ contains a copy of ℓ_∞.*

Proof. Sufficiency is obvious. Let us show necessity. Suppose that $L_p(\mu, X)$ contains a copy of ℓ_∞ and let T be an isomorphic embedding from ℓ_∞ into $L_p(\mu, X)$. By Lemma 1.6.1 we know that there exists a σ-finite and separable measure algebra $(\Omega_1, \Sigma_1, \mu_1) \subset (\Omega, \Sigma, \mu)$ such that

$$T(e_n) \in L_p(\mu_1, X)$$

for all $n \in \mathbb{N}$, where (e_n) denotes the canonical unit vector sequence in ℓ_∞. Let P be the projection operator from $L_p(\mu, X)$ onto $L_p(\mu_1, X)$ given in the preceding lemma. If we take $T_0 = P \circ T$ it is clear that,

$$\|T_0(e_n)\|_p = \|T(e_n)\|_p \not\to 0.$$

Then, by Theorem 1.3.1, $L_p(\mu_1, X)$ contains a copy of ℓ_∞. This completes the proof.

If we put together the preceding proposition and Proposition 1.6.1 we obtain

Theorem 1.6.2. *Let (Ω, Σ, μ) be an arbitrary positive measure space, let $1 \leq p < \infty$, and assume that $L_p(\mu, X)$ contains a copy of ℓ_∞. Then there exists a separable, finite and positive measure space $(\Omega_0, \Sigma_0, \mu_0)$ such that $L_p(\mu_0, X)$ contains a copy of ℓ_∞.*

Remark 1.6.3. In the preceding theorem no assertion about complementability of ℓ_∞ has been made. This is because such an assertion would have been superfluous. Recall that ℓ_∞ enjoys an important property: it is injective (see [35, page 71] or [93, page 105]). This means that ℓ_∞ is complemented in every Banach space containing it. In other words, $X \supset \ell_\infty$ if and only if $X \supset \ell_\infty$.
$$\quad\quad\quad\quad\quad\quad\quad\quad\quad\quad\quad\quad\quad\quad\quad\quad\quad\quad\quad\quad\quad\quad\quad (c)$$

When one considers sequences in $L_p(\mu, X)$ spaces, one has very often to deal with different subsets of Ω associated to these sequences. In most cases it is trivial to show that such subsets are measurable, just because they can easily put as countable unions and intersections of sets which are obviously measurable. Nevertheless, there are cases in which the situation is not so easy, and in our work we will find one of these situations. Given a basis (e_n) in a Banach space and a sequence (f_n) in $L_p(\mu, X)$ it is easy to see that the set

$$\{\omega \in \Omega : (f_n(\omega)) \text{ is a equivalent to } (e_n)\}$$

is measurable. However, can we also assure that the set

$$\{\omega \in \Omega : (f_n(\omega)) \text{ has a subsequence equivalent to } (e_n)\}$$

is measurable?

We will see that the answer is "yes". The key to prove this will be a beautiful old result about the so called Souslin operation , also called operation (A) (see [109, Chapter 1, Section 7], [118, Chapter 2, Section 5]). This "operation" is defined as follows:

Definition 1.6.1. *Given a family* $(A_{n_1,\ldots,n_k})$ *of subsets of a certain set* Ω, *where* k *runs* $\mathbb{N}$ *and* $(n_1, \ldots, n_k)$ *belongs to* $\mathbb{N}^k$, *we say that the set*

$$\bigcup_{n_1, n_2, \ldots} \bigcap_{k=1}^{\infty} A_{n_1, \ldots, n_k}$$

where the union is over all infinite sequences of natural numbers, is obtained from application to the Souslin operation to the family $(A_{n_1,\ldots,n_k})$.

Our interest in Souslin operation lies in the following important property:

Theorem 1.6.3 (1.7 of [109]). *Given a finite (and complete) measure space* (Ω, Σ, μ), *the Souslin operation applied to a family of measurable sets is a measurable set.*

Notice that the point is that although Souslin operation is applied to a countable family, it involves an union over the continuum many infinite sequences of natural numbers.

Now we can deal with the question posed above.

Proposition 1.6.3. *Let (e_n) be a basis of a Banach space and let (f_n) be a sequence of X-valued measurable functions defined on the finite measure space (Ω, Σ, μ). Then the set*

$$\{\omega \in \Omega : (f_n(\omega)) \text{ has a subsequence equivalent to } (e_n)\}$$

is measurable.

Proof. Given $M > 0$, we say that a sequence (x_n) is M-equivalent to (e_n) if

$$\frac{1}{M} \| \sum \lambda_n e_n \| \leq \| \sum \lambda_n x_n \| \leq M \| \sum \lambda_n e_n \|$$

for all finite sequences of scalars (λ_n). Take any finite sequence $n_1, \ldots, n_k$ of natural numbers. If $n_1 < n_2 < \ldots < n_k$ put as $A_{n_1,\ldots,n_k}$ the set of all $\omega \in \Omega$ such that

$$\frac{1}{M} \| \sum_{j=1}^{k} \lambda_j e_j \| \leq \| \sum_{j=1}^{k} \lambda_j f_{n_j}(\omega) \| \leq M \| \sum_{j=1}^{k} \lambda_j e_j \|$$

for all $(\lambda_j) \in \mathbb{Q}^k$. Otherwise, put $A_{n_1,\ldots,n_k} = \emptyset$. Of course all these sets are measurable, and so, by the preceding theorem, if we apply the Souslin operation to the family $(A_{n_1,\ldots,n_k})$ we get a measurable set. But notice that

$$\bigcup_{n_1, n_2, \ldots} \bigcap_{k=1}^{\infty} A_{n_1,\ldots,n_k} = \bigcup_{n_1 < n_2 < \ldots} \bigcap_{k=1}^{\infty} A_{n_1,\ldots,n_k}$$
$$= \{\omega \in \Omega : (f_n(\omega)) \text{ has a subsequence } M\text{-equivalent to } (e_n)\},$$

where the second union is over all infinite *increasing* sequences of natural numbers. Therefore, the following equality finishes the proof

$$\{\omega \in \Omega : (f_n(\omega)) \text{ has a subsequence equivalent to } (e_n)\}$$
$$= \bigcup_{m=1}^{\infty} \{\omega \in \Omega : (f_n(\omega)) \text{ has a subsequence } m\text{-equivalent to } (e_n)\}$$

As an immediate consequence we have the following

Corollary 1.6.1. *Let (Ω, Σ, μ) be an arbitrary measure space, let (e_n) be a basis of a certain Banach space and let (f_n) be a sequence in $L_p(\mu, X)$, where $1 \leq p < +\infty$. Then the set*

$$\{\omega \in \Omega : (f_n(\omega)) \text{ has a subsequence equivalent to } (e_n)\}$$

is measurable.

Proof. We use the standard reduction argument. By Lemma 1.6.1, there exists a σ-finite measure space $(\Omega_1, \Sigma_1, \mu_1) \subset (\Omega, \Sigma, \mu)$ such that (f_n) lies in $L_p(\mu_1, X)$. Now, by Proposition 1.6.1, there exists a probability measure μ_0 on (Ω_1, Σ_1) and an isometric isomorphism from $L_p(\mu_1, X)$ onto $L_p(\mu_0, X)$. Moreover, as we showed in the proof, this isometric isomorphism has the form

$$f \longrightarrow hf$$

where h is a certain scalar function which does not vanish at any point. So it is clear that for each $\omega \in \Omega_1$, $(h(\omega) f_n(\omega))$ has a subsequence equivalent to (e_n) if and only if so does $(f_n(\omega))$. Now the conclusion follows from the preceding proposition.

Now we wish to touch on another basic aspect of $L_p(\mu, X)$ spaces. Let us remember that for many purposes the most important scalar L_p spaces are ℓ_p and $L_p([0, 1])$. We will see that something similar happens also in $L_p(\mu, X)$ spaces. To understand well this, one should study the isometric theory of $L_p(\mu)$ spaces (see [84] and the Notes and Remarks of this Chapter), but we have preferred to give only a couple of fundamental results which will be needed in Chapters 4 and 5. We include simple and elementary proofs of them. The point is the following: when we have a finite measure μ, it is well known that *if it is purely atomic then the corresponding $L_p(\mu)$ space is isometrically isomorphic to ℓ_p, and otherwise, it has a complemented subspace isometrically isomorphic to $L_p([0, 1])$.* We will prove that the vectorial version of this result is also true.

If we take as $\Omega = \mathbb{N}$, Σ the σ-algebra of all subsets of $\mathbb{N}$, and μ the counting measure, that is, the measure defined by

$$\mu(A) = cardinal(A)$$

then, the corresponding $L_p(\mu, X)$ spaces are the known $\ell_p(X)$ spaces of all sequences (x_n) of vectors in X such that

$$\| (x_n) \|_p = \left(\sum_{n=1}^{\infty} \| x_n \|^p \right)^{\frac{1}{p}} < +\infty \quad \text{for } 1 \le p < \infty$$

$$\| (x_n) \|_\infty = \sup_n \| x_n \| < +\infty$$

with the $\| . \|_p$ norm. Moreover we have the following

Proposition 1.6.4. *Let $1 \le p \le +\infty$, if (Ω, Σ, μ) is a σ-finite purely atomic measure space then $L_p(\mu, X)$ is isometrically isomorphic to $\ell_p(X)$.*

Proof. Let (A_n) be a sequence of pairwise disjoint atoms of (Ω, Σ, μ) whose union is Ω. Then the maps

$$L_p(\mu, X) \quad \longrightarrow \quad \ell_p(X)$$

$$f = \sum_{n=1}^{\infty} \chi_{A_n}(.)x_n \quad \longrightarrow \quad (\mu(A_n)^{\frac{1}{p}}x_n)_n$$

for $1 \leq p < \infty$, and

$$L_\infty(\mu, X) \quad \longrightarrow \quad \ell_\infty(X)$$

$$f = \sum_{n=1}^{\infty} \chi_{A_n}(.)x_n \quad \longrightarrow \quad (x_n)_n$$

are isometric isomorphism onto.

Proposition 1.6.5. *If (Ω, Σ, μ) is not purely atomic and $1 \leq p \leq \infty$, then $L_p(\mu, X)$ contains a complemented copy of $L_p([0,1], X)$.*

Proof. Since (Ω, Σ, μ) is not purely atomic there exists a set $\Omega_1 \in \Sigma$ with finite and positive measure such that no subset of Ω_1 is an atom. Without loss of generality we may assume $\mu(\Omega_1) = 1$. Then we can find a Σ-partition of Ω_1 in two sets A_0 and A_1 of measure $\frac{1}{2}$. Now for $i = 0, 1$ we can find a Σ-partition of A_i in two sets A_{i0} and A_{i1} of measure $\frac{1}{4}$. Of course, in this way we construct a diadic decomposition $\{A_{i_1\ldots i_k} : i_1, \ldots, i_k \in \{0,1\}, k \in \mathbb{N}\}$ of Ω_1 in Σ-measurable sets, exactly in the analogous way to the classical diadic decomposition of $[0,1]$ in intervals $J_{i_1\ldots i_k}$ with left side in $0.i_1 \ldots i_k$ (the number written in binary form) and length $\frac{1}{2^k}$, with $k \in \mathbb{N}$. Let Σ_1 be the σ-algebra generated by the $A_{i_1\ldots i_k}$'s. Denoting by μ_1 the restriction of μ to Σ_1, Lemma 1.6.2 says that $L_p(\mu_1, X)$ is complemented in $L_p(\mu, X)$. So, to finish our proof it is enough to show that $L_p(\mu_1, X)$ is isometrically isomorphic to $L_p([0,1], X)$. But if we denote by S the subspace of $L_p(\mu_1, X)$ of all functions of the form $\sum \chi_{A_{i_1\ldots i_k}}(.)x_{i_1\ldots i_k}$, where we only consider finite sums and the $x_{i_1\ldots i_k}$'s are in X, it is very easy to show that the map

$$S \quad \rightarrow \quad L_p([0,1], X)$$

$$\sum \chi_{A_{i_1\ldots i_k}}(.)x_{i_1\ldots i_k} \quad \rightarrow \quad \sum \chi_{J_{i_1\ldots i_k}}(.)x_{i_1\ldots i_k}$$

is an isometry. Therefore, it can be extended to an isometric isomorphism from $L_p(\mu_1, X)$ onto $L_p([0,1], X)$.

To complete the section let us recall a well known concept which plays a central role in $L_p(\mu, X)$ spaces: *uniform integrability*. The subject deserves a lot of attention and we suggest the reading of Diestel's survey [36], which is entirely devoted to it. Here we only wish to give the definition and a couple of very important facts. To avoid useless difficulties we will assume that our measure space (Ω, Σ, μ) is finite, and so, for any $p > 1$ we have the continuous canonical embedding

$$i : L_p(\mu, X) \quad \hookrightarrow \quad L_1(\mu, X)$$
$$f \quad \rightarrow \quad f$$

Definition 1.6.2. *We say that a subset A of $L_1(\mu, X)$ is uniformly integrable if*

$$\lim_{\mu(E) \to 0} \int_E \|f\| \, d\mu = 0$$

uniformly in $f \in A$.

Given $p > 1$, $f \in L_p(\mu, X)$ and $E \in \Sigma$, it follows from Hölder inequality that

$$\int_E \|f\| \, d\mu = \int_\Omega \|f\| \, \chi_E \, d\mu \le \|f\|_p \, \mu(E)^{\frac{1}{q}}$$

where $\frac{1}{p} + \frac{1}{q} = 1$. As an immediate consequence we have that *subsets of $L_1(\mu, X)$ which are (contained and) bounded in $L_p(\mu, X)$ for some p, with $1 < p \le +\infty$, are uniformly integrable*. These are perhaps the simplest examples of uniformly integrable subsets of $L_1(\mu, X)$.

One of the most important applications of uniform integrability is Dunford-Pettis celebrated characterization of relatively weakly compact subsets of $L_1(\mu)$. It says that they are precisely the bounded and uniformly integrable sets [39, Theorem III.2.15] [51, Corollary IV.8.11.].

1.7 The dual of $C(K, X)$

In Section 1.5 we got a nice and useful representation of the dual of $L_p(\mu, X)$. What can we say about the dual of $C(K, X)$? The classical Riesz representation theorem establishes that the dual of $C(K)$ may be viewed as a space of measures via integration theory. Singer got the extension of this theorem to the vector-valued case.

Before stating Singer's result let us remember a few things about vector measures.

If K is a compact and Hausdorff space we denote by $\mathcal{B}(K)$ the σ-algebra of all Borel subsets of K. Given a finitely additive measure

$$m : \mathcal{B}(K) \to X^*$$

of bounded variation, for each X-valued $\mathcal{B}(K)$-simple function $f = \sum_i \chi_{A_i} x_i$, one can define the integral

$$\int_K f \, dm = \sum_i m(A_i)(x_i)$$

This integral is well defined and, using that m has bounded variation, it can be extended to all X-valued functions which are uniform limits of $\mathcal{B}(K)$-simple functions. In particular, the integral makes sense for all functions belonging to $C(K, X)$, and we get in this way a continuous linear functional on $C(K, X)$. Likewise in the scalar case, in general, many additive X^*-valued measures on $\mathcal{B}(K)$ with bounded variation coincide in $C(K, X)$ or, in other words, many of them define the same member of $C(K, X)^*$. However, only one of them is regular (a vector measure m is regular whenever its variation $\|m\|$ is regular). Let us denote by $rcabv(K, X^*)$ the closed subspace of $cabv(\mathcal{B}(K), X^*)$ of all its regular members. As in the scalar case it is also true that each functional in $C(K, X)^*$ can be represented by a measure in $rcabv(K, X^*)$, moreover $C(K, X)^*$ may be identified with $rcabv(K, X^*)$. This is Singer's extension of the classical Riesz theorem representing the dual of $C(K)$. It may be found in [125, Lemma 1.6, Section 1.4., Chapter II] (see also [70]).

Theorem 1.7.1 (Riesz-Singer). *The map*

$$rcabv(K, X^*) \longrightarrow C(K, X)^*$$
$$m \longrightarrow \tau_m$$

where τ_m *is defined by*

$$\tau_m(f) = \int_K f \, dm$$

for each $f \in C(K, X)$, *is an isometric isomorphism onto* $C(K, X)^*$.

It is worth to mention that the simplest members of $rcabv(K, X^*)$ are the linear combinations of measures of the form $\mu(.)x^*$, where μ is a scalar regular measure on K, and $x^* \in X^*$. In particular, we can take measures of the form $\delta_t(.)x^*$, with $t \in K$, where of course, for each $A \in \mathcal{B}(K)$

$$\delta_t(A) = \begin{cases} 1 & \text{if } t \in A \\ 0 & \text{otherwise} \end{cases}$$

Remark 1.7.1. If $m : \mathcal{B}(K) \to X^*$ is a finitely additive measure, for each $x \in X$, we can define the scalar measure m_x by

$$m_x(A) = m(A)(x)$$

for all $A \in \mathcal{B}(K)$. If m_x is countably additive for all $x \in X$, it is said that m is w^*-countably additive. Analogously, m is said to be w^*-regular if m_x is regular for all $x \in X$. It is easy to prove that if m is of bounded variation then m is w^*-countably additive if and only if it is countably additive, and m is w^*-regular if and only if it is regular. Moreover, thanks to Alexandroff theorem ([51, Theorem III.5.13]), each X^*-valued finitely additive w^*-regular measure of bounded variation defined on K belongs to $rcabv(K, X^*)$.

It is not difficult now to characterize weak convergence in $C(K, X)$. The following result is very well known.

Proposition 1.7.1 (Theorem 9, [42]). *Let (f_n) be a bounded sequence in $C(K, X)$, and $f \in C(K, X)$. Then (f_n) is weakly convergent to f if and only if $(f_n(t))$ is weakly convergent to $f(t)$ for each $t \in K$.*

Proof. Assume (f_n) is a sequence in $C(K, X)$ weakly convergent to $f \in C(K, X)$. Let t_0 be a point of K. For each $x^* \in X^*$ let us consider the measure $\delta_{t_0}(.)x^*$ as a linear form in $C(K, X)$. It follows that

$$\lim_n x^*(f_n(t_0)) = \lim_n \int_K f_n \, d(\delta_{t_0}(.)x^*) = \int_K f \, d(\delta_{t_0}(.)x^*) = x^*(f(t_0))$$

Therefore, $(f_n(t_0))$ is weakly convergent to $f(t_0)$.

For the converse, assume that (f_n) is a bounded sequence in $C(K, X)$ such that $(f_n(t))$ is weakly convergent to $f(t)$ for each $t \in K$. Let us take a continuous linear form on $C(K, X)$, that is, a measure m in $rcabv(K, X^*)$. Let us consider the variation $\|m\|$ of m. Extending $\|m\|$ in the usual way we can assume the measure $\|m\|$ is complete. Then theorem 1.5.2 implies that there exists $\varphi : K \to X^*$ bounded w^*-measurable such that for each $A \in \Sigma$ and all $x \in X$

$$m(A)(x) = \int_A <x, \varphi(t)> d\|m\|(t)$$

Of course, this means that

$$\int_K h \, dm = \int_K <h(t), \varphi(t)> d\|m\|(t) \tag{1.12}$$

for every X-valued $\mathcal{B}(K)$-simple function h defined on K, and this implies that 1.12 is also true for all continuous functions $h \in C(K, X)$. Now, it follows from the usual Lebesgue dominated convergence theorem that

$$\lim_n \int_K f_n \, dm = \lim_n \int_K <f_n(t), \varphi(t)> d\|m\|(t)$$
$$= \int_K <f(t), \varphi(t)> d\|m\|(t) = \int_K f \, dm$$

This completes the proof.

Next corollary follows immediately from the preceding proposition and the following well known observation: a sequence (x_n) in a Banach space X is weakly Cauchy if and only if given two subsequences (y_n) and (z_n) of (x_n), their difference $(y_n - z_n)$ is weakly null.

Corollary 1.7.1. *Let (f_n) be a bounded sequence in $C(K, X)$, then (f_n) is weakly Cauchy if and only if $(f_n(t))$ is weakly Cauchy for each $t \in K$.*

1.8 Notes and Remarks

The results of the first section are due to Bessaga and Pełczyński and are essentially contained in their classical work [4] of 1958.

In our work, Theorem 1.1.2 will be the key to find *complemented* copies of c_0. We believe that this Theorem has been "well known" for a long time (see for instance Exercise 8 of Chapter V of [35], which is a closely related result due to Pełczyński [99]) and it is in the spirit of the classical Bessaga-Pełczyński work. However we have not found explicitly stated in the literature the precise version we give. In any case, versions of this result may be found in [52] or [122].

The relatively recent paper of Rosenthal [114] must be consulted by anybody dealing with Banach spaces containing c_0. Although we will not use Rosenthal's paper, it is worth mentioning at least one of its main results ([114, Corollary 1.2]). In the statement the following definition is needed.

Definition. *A sequence (b_j) in a Banach space is called strongly summing, or, in short, (s.s.), if (b_j) is a weak-Cauchy basic sequence so that the series $\sum c_j$ of scalars converges whenever*

$$\sup_n \| \sum_{j=1}^{n} c_j b_j \| < +\infty$$

A weak-Cauchy sequence is called *non-trivial* if it is non-weakly convergent.

Theorem (Rosenthal, 1994, [114]). *c_0 does not embed in a Banach space X if and only if every non-trivial weak-Cauchy sequence in X has an (s.s.)-subsequence.*

It seems that this theorem, or some of the other results in [114], could have natural applications in the study of vector-valued function spaces, and in general in Banach space theory

The results of Sections 2 and 3 on copies of ℓ_1 and ℓ_∞ are mainly due to Rosenthal. The key result to find *complemented* copies of ℓ_1 is of course Theorem 1.2.1. It is well known and we believe that it is essentially contained in [110, 111], but we have not found it explicitly in the literature either. However, we must call the attention to a result in the literature which is actually an improvement of Theorem 1.2.1: Bourgain in [18, Proposition 11 of Appendix I] provides a characterization of sequences which are *finite* union of complemented ℓ_1-sequences.

Proposition 1.3.1 and Corollary 1.3.3 are important to get later the results about copies of ℓ_∞ in the spaces we are interested in. They were obtained by Drewnowski, but, as he points out, they have their roots in the celebrated work of Kalton on spaces of compact operators [78, Proposition 4].

In Section 1.5 we give two old Radon-Nikodým theorems and two theorems giving a representation of the dual of $L_p(\mu, X)$ spaces. All these results are due to A. and C. Ionescu Tulcea and follow from the theory of liftings.

Lifting theory seems a little bit old fashioned, at least for the general analyst. Many of us feel unease when we need consult the two main monographs which consider extensively the subject: Dinculeanu's "Vector Measures" [40] and A. and C. Ionescu Tulcea's "Topics in the Theory of Liftings" [75] (both written in the sixties). We may add that Diestel and Uhl, in their monograph "Vector Measures" [39], seem skeptical about the usefulness of abstract representations of functionals and operators on $L_p(\mu, X)$ using lifting theory. For instance, they say in the Notes and Remarks of Chapter IV: "... *neither of which* (they mean the abstract representations of functionals on $L_p(\mu, X)$) *has found concrete applications to the structure of $L_p(\mu, X)$ as yet*". (see also their Notes and Remarks of Chapter III).

Diestel and Uhl's point of view was right in the seventies, but now, applications of lifting theory to vector measures and to the representation of the dual of $L_p(\mu, X)$ have already shown their usefulness and have to be used for any dealing with our subject. We can mention at least the following points in which, as far as we know, this theory is essential:

1. E. and P. Saab's theorem (theorem 3.1.4) on complemented copies of ℓ_1 in $C(K, X)$.
2. The Talagrand's fundamental paper on weak Cauchy sequences in $L_1(E)$ [127].
3. Theorem 4.1.2 on complemented copies of ℓ_1 in $L_p(\mu, X)$ (it uses the representation of the dual of $L_p(\mu, X)$ and it also uses E. and P. Saab's theorem of point 1).

Although the results of this section in their greatest generality are due to A. and C. Ionescu Tulcea, they have a long history. Many authors contributed to get those results, proving them in more and more general conditions. In particular, for separable Banach spaces X the representation results of the section are much easier and do not need lifting theory at all. We believe that it is interesting to know how the situation is in this important particular case.

Notice first that in the whole section we have been working with functions instead of equivalence classes of functions, and this have been troublesome. However, we have to do that because we need the functions $\| \Psi(.) \|$ to be measurable for our representing functions Ψ, and it is easy to find examples (in *non* separable spaces X) showing that this is not true for all functions Ψ in $\mathcal{L}_{w^*}^{\infty}(\lambda, X^*)$. So we have to be careful about which representing function we take among all possible choices. Or, in other words, if we consider equivalence classes of functions in $\mathcal{L}_{w^*}^{\infty}(\lambda, X^*)$ we have that not all functions of a given equivalence class enjoy the aforementioned nice measurability property.

When X is separable the preceding measurability problem disappears. It is easy to show that in this case the functions $\| \Psi(.) \|$ are measurable for all members Ψ of $\mathcal{L}_{w^*}^{p}(\lambda, X^*)$, for $1 \leq p \leq +\infty$. Moreover, the equivalence relation $\stackrel{*}{\sim}$ is just the usual "to be equal almost everywhere" equivalence relation. So, when X is separable everything works fine identifying functions with equivalence classes of functions as usual.

On the other hand, for separable Banach spaces X the key result Theorem 1.5.2 was already proved in the classical Dunford and Pettis' paper [50] of 1940 (see theorem 2.1.1 and in general all Chapter II, Part 1, Section A of [50]). The techniques they used can be found in [51, Chapter VI, Section 8]. Observe that this Theorem 1.5.2 was the only one in the section in which lifting theory is used. In fact we deduce in quite a standard way the rest of representation results from it. A very general representation of the dual of $L_p(\mu, X)$ spaces for separable spaces X was given by Dieudonné [41] in 1951.

For simplicity we have been working in Section 1.5 in finite measure spaces because this is the most important case, and it is the case we will need in our work. However, it should be remarked that all the results are true in much more general conditions (see [40, 75]). In particular, it is very easy to extend the results to σ-finite measure spaces. We could add that in the case $1 < p < +\infty$, the representation theorems of the dual of $L_p(\mu, X)$ spaces are true for arbitrary positive measure spaces (this has been noticed at least by Hu and Lin in [74]).

The first aim of Section 1.6 is to show that, for $1 \leq p < +\infty$, our study of $L_p(\mu, X)$ spaces may be reduced to the finite measure case. It should be pointed out that we do not get a reduction for $p = +\infty$. This case seems to be completely different and will be considered in Chapter 5.

We make first a reduction to the σ-finite case, and once this is done, we simply notice that every $L_p(\mu, X)$ space with σ-finite measure is isometrically isomorphic to an $L_p(\mu_0, X)$ space with finite measure μ_0 (Proposition 1.6.1). We would like to recall another approach. Given a σ-finite measure space (Ω, Σ, μ) there exists a sequence (A_n) of pairwise disjoint measurable sets of finite and positive measure whose union is Ω. Let us call μ_n the restriction of μ to $\Sigma(A_n) = \{A \in \Sigma : A \subset A_n\}$. Of course, the μ_n's are finite measures. On the other hand, it is clear that $L_p(\mu, X)$ is isometrically isomorphic to $(\sum \oplus L_p(\mu_n, X))_p$ for $1 \leq p \leq \infty$, where, as usual, $(\sum \oplus X_n)_p$ denotes the sum of the Banach spaces X_n in the sense of ℓ_p (see [93]). In this way, it would not be difficult to show, for instance, that if (f_k) is a c_0-sequence (respectively an ℓ_1-sequence) in $L_p(\mu, X)$, then there exists $n \in \mathbb{N}$ such that $(\chi_{A_n}(.)f_k)$ has a c_0-subsequence (respectively, an ℓ_1-subsequence) in the space $L_p(\mu_n, X)$.

The second goal of Section 1.6 is a problem of measurability (proposition 1.6.3 and corollary 1.6.1). Probably these results are well known but we have not been able to find them in the literature. We believe that the application of Souslin operation we give could be useful in some other similar situations.

The third goal of the section is showing how the differences between $L_p(\mu)$ spaces depending on the measure μ is purely atomic or not, have a natural immediate translation into $L_p(\mu, X)$ spaces. Actually, the natural place of the results in this direction (Propositions 1.6.4 and 1.6.5) would be the isometric theory of L_p spaces (see [84, Chapter 5]). We have preferred to avoid its technical difficulties, but let us say a few words. Recall that the important thing in the definition of $L_p(\mu)$ (and in the same way, of $L_p(\mu, X)$) is not the measure space (Ω, Σ, μ), but the underlying measure algebra $(\tilde{\Sigma}, \tilde{\mu})$. In other words, given two measure spaces $(\Omega_1, \Sigma_1, \mu_1)$ and $(\Omega_2, \Sigma_2, \mu_2)$, $L_p(\mu_1, X)$ and $L_p(\mu_2, X)$ are isometrically isomorphic whenever the corresponding measure algebras $(\tilde{\Sigma}_1, \tilde{\mu}_1)$ and $(\tilde{\Sigma}_2, \tilde{\mu}_2)$ are isometric. In particular, we deduce from Caratheodory Theorem [84, Section 14, Theorem 5 ,and Corollary to Theorem 9] that if (Ω, Σ, μ) is a separable (and σ-finite) measure space then $L_p(\mu, X)$ is isometrically isomorphic to one of the following spaces:

1. $L_p([0,1], X)$ (purely nonatomic case),
2. $\ell_p(X)$ (purely atomic case),
3. $(L_p([0,1], X) \oplus \ell_p(X))_p$ (non purely atomic case, infinite number of atoms),
4. $(L_p([0,1], X) \oplus \ell_p^m(X))_p$ (non purely atomic case, m atoms),

where $\ell_p^m(X)$ denotes the sum of X in the sense of ℓ_p, m times.

Last paragraphs of the section are devoted to the important concept of *uniform integrability* and we suggest the reading of the survey paper [36]. There is a minor technical difference between the definition of [36] and the one given here (see Theorem 1 of [36]), but this should not cause any problem.

Finally, Section 1.7 deals with $C(K, X)$ spaces. We give the classical extension of Riesz theorem found by Singer, and the characterization of weak convergence in $C(K, X)$ (proposition 1.7.1). This last result has been very well know for a long time but, as far as we know, it was noticed for the first time in its complete generality by Dobrakov in [42, Theorem 9].

2. Copies of c_0 and ℓ_1 in $L_p(\mu, X)$

We can say that the starting point in the search of copies of c_0 and ℓ_1 in vector-valued function spaces is in the pioneering works of Hoffmann-Jørgensen [73], Kwapień [83] and Pisier [101] in the seventies. The results obtained then were the following:

Theorem (Kwapień, 1974 [83]**).** *For $1 \leq p < \infty$, $L_p(\mu, X)$ contains a copy of c_0 if and only if X does.*

Theorem (Pisier, 1978 [101]**).** *For $1 < p < \infty$, $L_p(\mu, X)$ contains a copy of ℓ_1 if and only if X does.*

It should be mentioned that Kwapień's theorem lies very much on the Hoffmann-Jørgensen's preliminary work [73].

This chapter is devoted to study these two theorems, but in our way we will have occasion of learning also some other interesting results such as Bourgain's theorem 2.1.1 on averaging of c_0-sequences, Kadec-Pełczyński-Rosenthal lemma 2.1.3, or Maurey-Pisier-Bourgain theorem 2.2.1 on ℓ_1-sequences in $L_p(\mu, X)$.

Since we are interested in the search of copies of c_0 and ℓ_1 in $L_p(\mu, X)$, let us remember first which is the situation in the scalar case. We can make the following quite elementary analysis. For $1 < p < \infty$, $L_p(\mu)$ spaces are reflexive, and so it is clear that they can not contain copies of any of the spaces we are considering: c_0, ℓ_1, or ℓ_∞. It is also clear that $L_1(\mu)$ does not contain copies of c_0, because $L_1(\mu)$ is weakly sequentially complete and c_0 does not enjoy this property. Of course, if it does not contain copies of c_0, it can not contain the bigger space ℓ_∞.

Notice that an idea behind Kwapień's theorem is the following: "Since $L_p(\mu)$ has no copies of c_0, if c_0 lies in $L_p(\mu, X)$ then it must lie in X". And a similar philosofy is underlying Pisier's theorem, too. As we mentioned in the introduction these ideas have been very important and fruitful in the study of vector-valued function Banach spaces.

Soon after the works of Hoffmann-Jørgensen, Kwapień and Pisier, Bourgain [14, 16] provided new proofs of the preceding theorems, giving results and introducing points of view which turned out to be very important in subsequent works. This is why Bourgain's ideas lie all over this chapter. In particular, we will follow his approach to get Kwapień's theorem because it contains a result (theorem 2.1.1) which will be crucial in Chapter 3 to get Saab and Saab's theorem 3.1.4.

2.1 Copies of c_0 in $L_p(\mu, X)$

If one thinks a little bit how to show that if $L_p(\mu, X)$ contains a copy of c_0 then X must contain a copy of c_0, too, a first natural question arises: If (f_n) is a c_0-sequence in $L_p(\mu, X)$, can we take $\omega_0 \in \Omega$ such that $(f_n(\omega_0))$ is a c_0-sequence in X? However, one immediately realizes that this question is not properly formulated, since to find a point in a measure space is in general meaningless, because singletons are very often null sets. On the other hand, trying to make such an assertion about the *sequence* $(f_n(\omega_0))$ itself is clearly too much, even in the simplest situation. When we have a c_0-sequence (x_n, y_n) in $X \times X$, for instance in $c_0 \times c_0$, we can not hope that either (x_n) or (y_n) to be a c_0-sequence, we only know that either (x_n) or (y_n) have a c_0-*subsequence*. Of course, this can be easily translated to $L_p(\mu, X)$. We can give for instance the following quite trivial example: Let A_1, A_2 be two disjoint measurable sets with finite positive measure, and let (e_n) be the canonical basis of c_0, then we define

$$f_n = \chi_{A_i}(.)e_n$$

for each $n \in \mathbb{N}$, where $i = 1$ if n is odd and $i = 2$, otherwise. It is immediate that (f_n) is a c_0-sequence in $L_p(\mu, c_0)$, but $(f_n(\omega))$ is not a c_0-sequence for any point $\omega \in \Omega$.

The preceding arguments lead to a more accurate formulation of the question: if (f_n) is a c_0-sequence in $L_p(\mu, X)$, can we find a subsequence (f_{n_k}) of (f_n) and measurable subset A of Ω, with positive measure, such that $(f_{n_k}(\omega))$ is a c_0-sequence in X for each $\omega \in A$? The following simple example shows that the answer is "No".

Example 2.1.1. For $1 \leq p < +\infty$ there is a c_0-sequence (f_n) in $L_p([0,1], c_0)$ such that for each subsequence (f_{n_k}) of (f_n), $(f_{n_k}(t))$ is not a c_0-sequence for (almost) any $t \in [0,1]$.

It is enough to define for each $n \in \mathbb{N}$

$$f_n : [0,1] \longrightarrow c_0$$
$$t \longmapsto (r_n(t) + 1)e_n$$

where (e_n) denotes the canonical basis of c_0, and (r_n) is the sequence of Rademacher functions (remember that Rademacher functions (r_n) are defined by

$$r_n(t) = sign(sin(2^n \pi t))$$

for all $t \in [0,1]$ and all $n \in \mathbb{N}$).

On one hand, it is clear that for each subsequence (f_{n_k}) of (f_n), and for almost all $t \in [0,1]$ the sequence $(f_{n_k}(t))$ vanishes for infinite natural numbers k, and so it can not be a c_0-sequence. On the other hand, let us show that (f_n) is a c_0-sequence in $L_p([0,1], c_0)$. Let (λ_n) be a finite sequence of scalars. We have

$$
\begin{aligned}
\max_n |\lambda_n| &= \max_n |\lambda_n| \int_0^1 |r_n(t) + 1| \, dt \\
&= \max_n \int_0^1 |\lambda_n(r_n(t) + 1)| \, dt \\
&\leq \int_0^1 \max_n |\lambda_n(r_n(t) + 1)| \, dt \\
&= \int_0^1 \| \sum_n \lambda_n f_n(t) \| \, dt \\
&= \| \sum_n \lambda_n f_n \|_1 \\
&\leq \| \sum_n \lambda_n f_n \|_p .
\end{aligned}
$$

And we also have

$$
\begin{aligned}
\| \sum_n \lambda_n f_n \|_p &= (\int_0^1 \| \sum_n \lambda_n f_n(t) \|^p \, dt)^{1/p} \\
&= (\int_0^1 (\max_n |\lambda_n(r_n(t) + 1)|)^p dt)^{1/p} \\
&\leq (\int_0^1 (\max_n 2 |\lambda_n|)^p dt)^{1/p} \\
&= 2 \max_n |\lambda_n| .
\end{aligned}
$$

Therefore (f_n) is indeed a c_0-sequence in $L_p([0,1], c_0)$.

If we look at the sequence (f_n) of the preceding example we see that although $(f_{n_k}(t))$ is not a c_0-sequence for (almost) any $t \in [0,1]$, $(f_n(t))$ *does* have a c_0-*subsequence* for (almost) all $t \in [0,1]$ (the point is that the subsequence depends on the t). This leads us to a new question: if (f_n) is a c_0-sequence in $L_p(\mu, X)$, can we assure that the set of all $\omega \in \Omega$ for which $(f_n(\omega))$ has a c_0-*subsequence* is non null? Bourgain showed that the answer is "Yes" in the case $p = 1$, and we will see that for $1 < p < \infty$ the answer is "Yes", too. Of course, this provides an answer to the problem posed at the beginning of the section: this guarantees that if $L_p(\mu, X)$ contains a copy of c_0 then X has the same property.

It might be surprising that some results in this section are stated for *seminormed* spaces. However this is not a more or less superfluous search of generality. We will see in Theorem 2.1.2 and in Saab and Saab's theorem 3.1.4 that this is actually convenient for our purposes. We will consider c_0-sequences in seminormed spaces. Denote by c_{00} the normed subspace of c_0 of all sequences of only finitely many non-zero terms. We will say that a sequence (x_n) in a seminormed space $(X, \| \cdot \|)$ is a c_0-sequence if there exist two positive numbers M, δ such that

$$\delta \sup \mid a_n \mid \ \leq \ \| \sum a_n x_n \| \ \leq \ M \sup \mid a_n \mid$$

for all $(a_n) \in c_{00}$. Of course this simply means that, by passing to the quotient normed space, the corresponding sequence is a c_0-sequence in (the completion of) this normed space, too. In the proofs of the few results we will state for seminormed spaces the only facts of the theory of normed spaces which are applied are Bessaga-Pełczyński theorem 1.1.1 and Bessaga-Pełczyński selection principle. It is easy to show that the usual proofs of these theorems also work for seminormed spaces or, better than this, one can see in the standard way, by passing to the quotient normed space, that Bessaga-Pełczyński results for seminormed spaces follow immediately from the corresponding ones in normed spaces.

Before going into Bourgain's approach we will summarize some important properties of Rademacher functions (r_n) in $[0, 1]$, because they will play a very important role in our study. These properties are of course quite well known, but for the sake of completeness we will include proofs. Proposition 2.1.2 is just a version of the "contraction principle" [37, 12.2].

Proposition 2.1.1. *Let $(X, \| \cdot \|)$ be a seminormed space. If $x, x_1, x_2, \ldots, x_m$ are elements of X, then*

$$\int_0^1 \| x + \sum_{i=1}^m r_{n_i}(t)x_i \| \, dt = \frac{1}{2^m} \sum_{\theta_i = \pm 1} \| x + \sum_{i=1}^m \theta_i x_i \|$$

for any collection of natural numbers $n_1 < n_2 < \ldots < n_m$. Consequently,

$$\int_0^1 \| x + \sum_{i=1}^m r_{n_i}(t)x_i \| \, dt = \int_0^1 \| x + \sum_{i=1}^m r_i(t)x_i \| \, dt.$$

Proof. The result is an immediate consequence of the fact that $r_{n_1}, r_{n_2}, \ldots, r_{n_m}$ are independent (symmetric) random variables which take the values 1 and -1 with probability $\frac{1}{2}$. In fact, if for each $\Theta = (\theta_1, \ldots, \theta_m) \in \{-1, 1\}^m$ we put

$$\Delta_\Theta = \{t \in [0, 1] : r_{n_i}(t) = \theta_i \text{ for } 1 \leq i \leq m\} = \bigcap_{i=1}^m \{t \in [0, 1] : r_{n_i}(t) = \theta_i\},$$

it is clear that this set has measure $\frac{1}{2^m}$. Since, for a given m, the Δ_Θ's form a partition of $[0, 1]$, we have

$$\int_0^1 \| x + \sum_{i=1}^m r_{n_i}(t)x_i \| dt = \sum_{\Theta \in \{-1,1\}^m} \int_{\Delta_\Theta} \| x + \sum_{i=1}^m r_{n_i}(t)x_i \| dt$$

$$= \frac{1}{2^m} \sum_{\theta_i = \pm 1} \| x + \sum_{i=1}^m \theta_i x_i \| .$$

Proposition 2.1.2. *Let* $(X, \| \cdot \|)$ *be a seminormed space. If* $x_0, x_1, x_2, \ldots, x_m$ *are elements of* X *and* $a_0, a_1, a_2, \ldots, a_m$ *are real numbers, then*

$$\int_0^1 \| a_0 x_0 + \sum_{i=1}^m a_i r_i(t)x_i \| dt \leq \max_{0 \leq i \leq m} | a_i | \int_0^1 \| x_0 + \sum_{i=1}^m r_i(t)x_i \| dt.$$

In particular

$$\| x_0 \| \leq \int_0^1 \| x_0 + \sum_{i=1}^m r_i(t)x_i \| dt, \tag{2.1}$$

and

$$\int_0^1 \| \sum_{i=1}^{m-1} r_i(t)x_i \| dt \leq \int_0^1 \| \sum_{i=1}^m r_i(t)x_i \| dt.$$

Proof. By homogeneity, we may assume $| a_i | \leq 1$ for $i \in \{0, 1, \ldots, m\}$.

Let us consider the real function defined on $B(\ell_\infty^{m+1})$, the unit ball of ℓ_∞^{m+1}, by

$$(a_0, a_1, \ldots, a_m) \longmapsto \int_0^1 \| a_0 x_0 + \sum_{i=1}^m a_i r_i(t)x_i \| dt$$

This function is obviously continuous and convex. Therefore, it takes its maximum among the extreme points of the compact set $B(\ell_\infty^{m+1})$, which are the points of the form $(\epsilon_0, \epsilon_1, \ldots, \epsilon_m)$ with $\epsilon_i = \pm 1$ for $i = 0, 1, \ldots, m$. Notice now that it follows easily from the preceding proposition that in all these points the function takes the same value. To be precise,

$$\int_0^1 \| \epsilon_0 x_0 + \sum_{i=1}^m \epsilon_i r_i(t)x_i \| dt = \int_0^1 \| x_0 + \sum_{i=1}^m r_i(t)x_i \| dt$$

for any choice of signs $\epsilon_0, \epsilon_1, \ldots, \epsilon_m = \pm 1$. Hence, for any $(a_0, a_1, \ldots, a_m) \in B(\ell_\infty^{m+1})$ we have

$$\int_0^1 \| a_0 x_0 + \sum_{i=1}^m a_i r_i(t)x_i \| dt \leq \int_0^1 \| x_0 + \sum_{i=1}^m r_i(t)x_i \| dt$$

This completes the proof of the first inequality. The others are just particular cases of it.

Now, let us begin to study Bourgain's approach. The key of this approach is Lemma 2.1.2, which is an "averaging version" of Bessaga-Pełczyński classical theorem 1.1.1 on c_0-sequences. In its proof we will use the following simple fact about sequences of positive numbers.

Lemma 2.1.1. Let (λ_i) and (δ_m) be sequences of positive numbers and assume that (λ_i) is bounded and non decreasing, then (λ_i) has a subsequence (λ_{i_m}) such that

$$\lambda_{i_{m+1}} \leq (1 + \delta_m)\lambda_{i_m}$$

for all $m \in \mathbb{N}$.

Proof. Observe first that the sequence (λ_i) must be convergent. Let us denote by λ its limit. Consider the sequence $(\frac{\lambda}{\lambda_i})$. It is clear that

$$\lim_i \frac{\lambda}{\lambda_i} = 1 < 1 + \delta_m$$

for all $m \in \mathbb{N}$. So, (λ_i) has a subsequence (λ_{i_m}) such that

$$\frac{\lambda}{\lambda_{i_m}} < 1 + \delta_m$$

for all $m \in \mathbb{N}$. Since (λ_i) is non decreasing, we have $\lambda_i \leq \lambda$ for all i. Therefore, in particular, we have

$$\frac{\lambda_{i_{m+1}}}{\lambda_{i_m}} \leq 1 + \delta_m$$

for all m.

Lemma 2.1.2. If $(X, \| \cdot \|)$ is a seminormed space and (x_i) is a non null sequence in X such that

$$\sup_n \int_0^1 \| \sum_{i=1}^n r_i(t)x_i \| \, dt < +\infty,$$

then (x_i) has a c_0-subsequence.

Proof. Let (x_i) be a sequence in X such that

$$\limsup \| x_i \| > 0 \quad \text{and} \quad \sup_n \int_0^1 \| \sum_{i=1}^n r_i(t)x_i \| \, dt = B < +\infty.$$

Observe first that according to Propositions 2.1.1 and 2.1.2, for each subsequence (x_{k_i}) of (x_k), we have

$$\sup_n \int_0^1 \| \sum_{i=1}^n r_i(t) x_{k_i} \| \, dt \;=\; \sup_n \int_0^1 \| \sum_{i=1}^n r_{k_i}(t) x_{k_i} \| \, dt$$

$$\leq \; \sup_m \int_0^1 \| \sum_{i=1}^m r_i(t) x_i \| \, dt = B < +\infty,$$

and therefore,

$$\sup_n \int_0^1 \| \sum_{i=1}^n r_i(t) x_{k_i} \| \, dt \leq B < +\infty \tag{2.2}$$

Then we can assume without loss of generality that there exists $\gamma > 0$ such that $\| x_i \| \geq \gamma > 0$ for all $i \in \mathbb{N}$. Let us see now that (x_i) must be weakly null. Given $x^* \in X^*$, thanks to the Khinchine's inequalities, for a certain $A > 0$ we have

$$A(\sum_{i=1}^m |<x_i, x^*>|^2)^{1/2} \;\leq\; \int_0^1 | \sum_{i=1}^m <x_i, x^*> r_i(t) | \, dt$$

$$= \; \int_0^1 |< \sum_{i=1}^m r_i(t) x_i, x^* >| \, dt$$

$$\leq \; \int_0^1 \| \sum_{i=1}^m r_i(t) x_i \| \, \| x^* \| \, dt \leq B \, \| x^* \|$$

for all $m \in \mathbb{N}$. Thus, $\sum_{i=1}^\infty |<x_i, x^*>|^2$ converges and so

$$\lim_i <x_i, x^*> = 0.$$

Take now a non increasing sequence (δ_n) of positive numbers such that

$$\prod_{n=1}^\infty (1 + \delta_n) < +\infty \quad \text{and} \quad \sum_{n=1}^\infty 2^n \delta_n < +\infty$$

Using the Bessaga-Pełczyński selection principle (see [35, page 42] or [93, pages 4-5]), we can extract a subsequence (y_i) of (x_i) in such a way that (y_i) is a basic sequence and satisfies

$$\| \sum_{i=1}^n a_i y_i \| \leq (1 + \delta_n) \| \sum_{i=1}^m a_i y_i \| \tag{2.3}$$

if $n < m$ and $a_1, a_2, \ldots, a_m$ are scalars.

By 2.2 and Proposition 2.1.2, $(\int_0^1 \| \sum_{j=1}^i r_j(t) y_j \| \, dt)_{i=1}^\infty$ is a bounded non decreasing sequence of positive numbers, so, it follows from lemma 2.1.1 that there is an increasing sequence (i_m) on natural numbers such that

$$\int_0^1 \| \sum_{i=1}^{i_{m+1}} r_i(t)y_i \| \, dt \le (1 + \delta_m) \int_0^1 \| \sum_{i=1}^{i_m} r_i(t)y_i \| \, dt \qquad (2.4)$$

for all $m \in \mathbb{N}$. Our aim now is to show that (y_{i_m}) is a c_0-sequence. This is the only point which remains to be proved, however it is the hardest part in the proof. Since (y_{i_m}) is a seminormalized basic sequence, by Theorem 1.1.1, we only need to show that $\sum y_{i_m}$ is a w.u.C. series, that is, we have to show that

$$\sup\{\| \sum_{n=1}^{m} \theta_n y_{i_n} \|: m \in \mathbb{N}, \theta_n = \pm 1\} < +\infty. \qquad (2.5)$$

To prove this let us define some quantities which relate the expressions involved in 2.3 and 2.4. For any finite choice of signs $(\theta_1, \theta_2, \ldots, \theta_m)$ put

$$\rho(\theta_1, \theta_2, \ldots, \theta_m) = \int_0^1 \| \sum_{n=1}^{m} \theta_n y_{i_n} + \sum_{i \in \sigma_m} r_i(t)y_i \| \, dt$$

where $\sigma_m = \{1, 2, 3, \ldots, i_m\} \setminus \{i_1, i_2, \ldots, i_m\}$. First of all, observe that it follows from 2.1 of Proposition 2.1.2 that

$$\| \sum_{n=1}^{m} \theta_n y_{i_n} \| \le \rho(\theta_1, \theta_2, \ldots, \theta_m)$$

Therefore, in order to prove 2.5, it is enough to show that the numbers $\rho(\theta_1, \theta_2, \ldots, \theta_m)$ are bounded. Second, notice that it follows immediately from 2.3 that for each choice of signs $(\theta_1, \theta_2, \ldots, \theta_m)$ we have

$$\begin{aligned}
\rho(\theta_1, \theta_2, \ldots, \theta_m) &= \int_0^1 \| \sum_{n=1}^{m} \theta_n y_{i_n} + \sum_{i \in \sigma_m} r_i(t)y_i \| \, dt \\
&\le \int_0^1 (1 + \delta_{i_m}) \| \sum_{n=1}^{m+1} \theta_n y_{i_n} + \sum_{i \in \sigma_{m+1}} r_i(t)y_i \| \, dt \\
&\le (1 + \delta_m)\rho(\theta_1, \ldots, \theta_m, \theta_{m+1}).
\end{aligned}$$

Therefore,

$$\rho(\theta_1, \theta_2, \ldots, \theta_m) \le (1 + \delta_m)\rho(\theta_1, \ldots, \theta_m, \theta_{m+1}) \qquad (2.6)$$

We will need another property of the numbers $\rho(\theta_1, \theta_2, \ldots, \theta_m)$ which is contained in the following Claim. We will prove it at the end of this proof.

Claim: $\int_0^1 \| \sum_{i=1}^{i_m} r_i(t)y_i \| \, dt = \int_0^1 \rho(r_1(t), r_2(t), \ldots, r_m(t)) \, dt$

With these properties of the $\rho(\theta_1, \theta_2, \ldots, \theta_m)$'s in mind we only have to do some computations to finish our proof. By our Claim and 2.4 we have

$$\int_0^1 \rho(r_1(t), \ldots, r_m(t), r_{m+1}(t))dt \;=\; \int_0^1 \Big\| \sum_{i=1}^{i_{m+1}} r_i(t)y_i \Big\| \, dt$$

$$\leq\; (1+\delta_m)\int_0^1 \Big\| \sum_{i=1}^{i_m} r_i(t)y_i \Big\| \, dt$$

$$=\; (1+\delta_m)\int_0^1 \rho(r_1(t), \ldots, r_m(t))dt,$$

and thus

$$\int_0^1 \left(\rho(r_1(t), \ldots, r_m(t), r_{m+1}(t)) - (1+\delta_m)\rho(r_1(t), \ldots, r_m(t)) \right) \, dt \leq 0.$$

Then, given a finite collection $(\theta_1, \ldots, \theta_m, \theta_{m+1})$ of signs and putting

$$S = \{t \in [0,1] : (r_1(t), \ldots, r_m(t), r_{m+1}(t)) \neq (\theta_1, \ldots, \theta_m, \theta_{m+1})\},$$

we have

$$\int_0^1 \left(\rho(r_1(t), \ldots, r_m(t), r_{m+1}(t)) - (1+\delta_m)\rho(r_1(t), \ldots, r_m(t)) \right) \, dt$$

$$= \; \frac{1}{2^{m+1}}\left(\rho(\theta_1, \ldots, \theta_m, \theta_{m+1}) - (1+\delta_m)\rho(\theta_1, \ldots, \theta_m) \right)$$

$$+ \; \int_S \left(\rho(r_1(t), \ldots, r_m(t), r_{m+1}(t)) - (1+\delta_m)\rho(r_1(t), \ldots, r_m(t)) \right) \, dt$$

$$\leq \; 0.$$

Hence, using 2.2, 2.6 and the Claim, we have

$$\frac{1}{2^{m+1}}\left(\rho(\theta_1, \ldots, \theta_m, \theta_{m+1}) - (1+\delta_m)\rho(\theta_1, \ldots, \theta_m) \right)$$

$$\leq \; \int_S \left((1+\delta_m)\rho(r_1(t), \ldots, r_m(t)) - \rho(r_1(t), \ldots, r_m(t), r_{m+1}(t)) \right) \, dt$$

$$\leq \; \int_S \left((1+\delta_m)\rho(r_1(t), \ldots, r_m(t)) - \frac{1}{1+\delta_m}\rho(r_1(t), \ldots, r_m(t)) \right) \, dt$$

$$= \; (\delta_m + 1 - \frac{1}{1+\delta_m}) \int_S \rho(r_1(t), \ldots, r_m(t)) \, dt$$

$$\leq \; 2\delta_m \int_0^1 \rho(r_1(t), \ldots, r_m(t)) \, dt$$

$$= \; 2\delta_m \int_0^1 \Big\| \sum_{i=1}^{i_m} r_i(t)y_i \Big\| \, dt \leq 2\delta_m B.$$

Therefore,

$$\rho(\theta_1, \ldots, \theta_m, \theta_{m+1}) \leq (1+\delta_m)\rho(\theta_1, \ldots, \theta_m) + 4B2^m \delta_m$$

Now, by induction, it is easy to see that

$$\rho(\theta_1, \ldots, \theta_m, \theta_{m+1}) \leq \prod_{n=1}^{m}(1 + \delta_n)\rho(\theta_1) + 4B \left(\prod_{n=1}^{m}(1 + \delta_n)\right)\left(\sum_{n=1}^{m} 2^n \delta_n\right)$$

and then

$$\rho(\theta_1, \ldots, \theta_m)$$
$$\leq (\rho(1) + \rho(-1)) \prod_{n=1}^{\infty}(1 + \delta_n) + 4B \left(\prod_{n=1}^{\infty}(1 + \delta_n)\right)\left(\sum_{n=1}^{\infty} 2^n \delta_n\right) < +\infty$$

for all $m \in \mathbb{N}$ and all finite collection of signs $(\theta_1, \ldots, \theta_m)$.

Hence we have shown that the numbers $\rho(\theta_1, \theta_2, \ldots, \theta_m)$ are bounded. To finish the proof of the lemma it only remains to show the Claim.

Proof of the Claim: It is enough to observe that, using Proposition 2.1.2, we have

$$\int_0^1 \rho(r_1(t), r_2(t), \ldots, r_m(t)) \, dt$$

$$= \int_0^1 \left(\int_0^1 \| \sum_{n=1}^{m} r_n(t)y_{i_n} + \sum_{i \in \sigma_m} r_i(s)y_i \| \, ds\right) dt$$

$$= \int_0^1 \left(\frac{1}{2^{i_m - m}} \sum_{\epsilon_i = \pm 1, i \in \sigma_m} \| \sum_{n=1}^{m} r_n(t)y_{i_n} + \sum_{i \in \sigma_m} \epsilon_i y_i \|\right) dt$$

$$= \frac{1}{2^{i_m - m}} \frac{1}{2^m} \sum_{\epsilon_i = \pm 1, i \in \sigma_m} \sum_{\theta_n = \pm 1, 1 \leq n \leq m} \| \sum_{n=1}^{m} \theta_n y_{i_n} + \sum_{i \in \sigma_m} \epsilon_i y_i \|$$

$$= \frac{1}{2^{i_m}} \| \sum_{i=1}^{i_m} \epsilon_i y_i \|$$

$$= \int_0^1 \| \sum_{i=1}^{i_m} r_i(t)y_i \| \, dt.$$

This completes the proof of the Claim and therefore, the proof of the lemma, too.

It is interesting to observe that, since

$$\int_0^1 \| \sum_{i=1}^{n} r_i(t)x_i \| \, dt = \frac{1}{2^n} \sum_{\theta_i = \pm 1} \| \sum_{i=1}^{n} \theta_i x_i \|,$$

the preceding lemma says that if (x_n) is a non null sequence for which the set

$$\{\frac{1}{2^n} \sum_{\theta_i = \pm 1} \| \sum_{i=1}^{n} \theta_i x_i \| : n \in \mathbb{N}, \ \theta_i = \pm 1\}$$

of averages is bounded, then it has a subsequence (y_i) for which the set

$$\{\| \sum_{i=1}^{n} \theta_i y_i \| : n \in \mathbb{N}, \ \theta_i = \pm 1\}$$

is bounded.

Notice also that the preceding lemma provides the following particular version of one of the theorems we are trying to show (Theorem 2.1.2):

Let (x_i) be a sequence in X and assume that $(r_i(.)x_i)$ is a c_0-sequence in $L_1([0,1], X)$, then (x_i) has a c_0-subsequence.

We can finally state and prove Bourgain's theorem.

Theorem 2.1.1 (Bourgain [14], 1978). *Let (Ω, Σ, μ) be a finite measure space and suppose that for each $\omega \in \Omega$, $| . |_\omega$ is a seminorm on c_{00} such that $\omega \mapsto | x |_\omega$ is integrable on (Ω, Σ, μ) for each $x \in c_{00}$. Let us define the seminorm $\| . \|_0$ on c_{00} by*

$$\| x \|_0 = \int_\Omega | x |_\omega \, d\mu(\omega)$$

for every $x \in c_{00}$.

If (x_i) is a sequence in c_{00} which is a c_0-sequence for $\| . \|_0$, then there exists a positive measure set $A \in \Sigma$ such that (x_i) has a c_0-subsequence for $| . |_\omega$ whenever $\omega \in A$.

Proof. Assume that the hypothesis of the theorem holds. It is clear that thanks to the preceding lemma it is enough to show that the following is true:

Claim: *There exists a positive measure set $A \in \Sigma$ such that*

$$\limsup_n | x_i |_\omega > 0 \quad and \quad \sup_n \int_0^1 | \sum_{i=1}^{n} r_i(t)x_i |_\omega \, dt < +\infty$$

for all $\omega \in A$.

Let M and δ be two positive numbers such that

$$\delta \max | a_i | \leq \| \sum a_i x_i \|_0 = \int_\Omega | \sum a_i x_i |_\omega \, d\mu(\omega) \leq M \max | a_i |$$

for all $(a_i) \in c_{00}$. By our hypothesis, for each $i \in \mathbb{N}$, the function ϕ_i defined by $\phi_i(\omega) = |x_i|_\omega$ for $\omega \in \Omega$, belongs to $L_1(\mu)$. Moreover, (ϕ_i) is a seminormalized sequence in $L_1(\mu)$ because

$$\delta \leq \| x_i \|_0 = \int_\Omega |x_i|_\omega \, d\mu(\omega) = \int_\Omega |\phi_i(\omega)| \, d\mu(\omega) = \| \phi_i \|_1 \leq M$$

for all $i \in \mathbb{N}$.

To prove that our Claim is true we will show that

(a) (ϕ_i) is uniformly integrable,

(b) $\mu(\{\omega \in \Omega : \limsup_i \phi_i(\omega) > 0\}) > 0$, and

(c) $\sup_n \int_0^1 |\sum_{i=1}^n r_i(t)x_i|_\omega \, dt < +\infty$ for almost all $\omega \in \Omega$.

If (a) does not hold, then it is well known (see [35, pages 93,89]) that there exist a sequence (A_i) of pairwise disjoint measurable sets, a subsequence of (ϕ_i) which we will continue to denote in the same way, and an $\epsilon > 0$ such that

$$\int_{A_i} \phi_i(\omega) \, d\mu(\omega) > \epsilon$$

for all $i \in \mathbb{N}$. Now, Rosenthal's disjointification lemma 1.2.1 says that there is an increasing sequence (i_k) of natural numbers for which

$$\int_{\cup_{n \neq k} A_{i_n}} \phi_{i_k}(\omega) \, d\mu(\omega) < \frac{\epsilon}{2}$$

for all $k \in \mathbb{N}$. Now, we reach the following contradiction: for each $n \in \mathbb{N}$ we have

$$M \geq \left\| \sum_{k=1}^n x_{i_k} \right\|_0 = \int_\Omega \left| \sum_{k=1}^n x_{i_k} \right|_\omega \, d\mu(\omega) \geq \sum_{m=1}^n \int_{A_{i_m}} \left| \sum_{k=1}^n x_{i_k} \right|_\omega \, d\mu(\omega)$$

$$\geq \sum_{m=1}^n \left(\int_{A_{i_m}} |x_{i_m}|_\omega \, d\mu(\omega) - \int_{A_{i_m}} \left(\sum_{k=1, k \neq m}^n |x_{i_k}|_\omega \right) d\mu(\omega) \right)$$

$$= \sum_{m=1}^n \left(\int_{A_{i_m}} \phi_{i_m}(\omega) \, d\mu(\omega) - \int_{A_{i_m}} \left(\sum_{k=1, k \neq m}^n \phi_{i_k}(\omega) \right) d\mu(\omega) \right)$$

$$\geq n\epsilon - \sum_{m=1}^n \int_{A_{i_m}} \left(\sum_{k=1, k \neq m}^n \phi_{i_k}(\omega) \right) d\mu(\omega)$$

$$= n\epsilon - \sum_{k=1}^n \left(\sum_{m=1, m \neq k}^n \int_{A_{i_m}} \phi_{i_k}(\omega) \, d\mu(\omega) \right)$$

$$= n\epsilon - \sum_{k=1}^n \left(\int_{\cup_{m=1, m \neq k}^n A_{i_m}} \phi_{i_k}(\omega) \, d\mu(\omega) \right) > n\epsilon - n\frac{\epsilon}{2} = n\frac{\epsilon}{2}.$$

Thus (a) is proved.

Now, let us note that (b) is an immediate consequence of (a): If we suppose that $\mu(\{\omega \in \Omega : \limsup_i \phi_i(\omega) > 0\}) = 0$ then $\lim_i \phi_i(\omega) = 0$ for almost all $\omega \in \Omega$, because the functions ϕ_i are non negative. Hence Vitali's lemma [51, IV.10.9.] says us that $\lim_i \| \phi_i \|_1 = 0$, which is of course a contradiction.

In order to prove (c), for each $n \in \mathbb{N}$ put

$$\xi_n(\omega) = \int_0^1 | \sum_{i=1}^n r_i(t) x_i |_\omega \, dt$$

for all $\omega \in \Omega$. Clearly ξ_n belongs to $L_1(\mu)$ since, by proposition 2.1.1,

$$\xi_n(\omega) = \frac{1}{2^n} \sum_{\theta_i = \pm 1} | \sum_{i=1}^n \theta_i x_i |_\omega,$$

and by our hypothesis, each function $\omega \mapsto | \sum_{i=1}^n \theta_i x_i |_\omega$ belongs to $L_1(\mu)$. According to Proposition 2.1.2 it is also clear that (ξ_n) is an increasing sequence. On the other hand, since for each $n \in \mathbb{N}$

$$\int_\Omega \xi_n(\omega) \, d\mu(\omega) = \frac{1}{2^n} \sum_{\theta_i = \pm 1} \int_\Omega | \sum_{i=1}^n \theta_i x_i |_\omega \, d\mu(\omega)$$

$$= \frac{1}{2^n} \sum_{\theta_i = \pm 1} \| \sum_{i=1}^n \theta_i x_i \|_0 \leq M,$$

we deduce from the Lebesgue's monotone convergence theorem that the function $\omega \mapsto \sup_n \xi_n(\omega)$ is integrable, and therefore,

$$\sup_n \xi_n(\omega) = \sup_n \int_0^1 | \sum_{i=1}^n r_i(t) x_i |_\omega \, dt < +\infty$$

for almost all $\omega \in \Omega$. This finishes the proof of (c).

Once we have shown (a),(b) and (c), to complete our proof it is enough to observe that (b) and (c) imply that our claim is true.

Now, the following two results are immediate consequences.

Theorem 2.1.2. *Let (Ω, Σ, μ) be a finite measure space and let (f_n) be a c_0-sequence in $L_1(\mu, X)$. Then, the set of all $\omega \in \Omega$ such that $(f_n(\omega))$ has a c_0-subsequence is a measurable set with positive measure.*

Proof. Let (f_n) be a c_0-sequence in $L_1(\mu, X)$. Thanks to Proposition 1.6.3, we only have to show that the set of all $\omega \in \Omega$ such that $(f_n(\omega))$ has a c_0-subsequence is a non null set.

By our assumption, there are two positive numbers M and δ such that

$$\delta \max_i | a_i | \leq \| \sum_i a_i f_i \|_1 = \int_\Omega \| \sum_i a_i f_i(\omega) \| \, d\mu(\omega) \leq M \max_i | a_i |$$

for all $(a_i) \in c_{00}$. This suggests to define for each $\omega \in \Omega$

$$| x |_\omega = \| \sum_i a_i f_i(\omega) \|$$

for all $x = \sum a_i e_i \in c_{00}$. Of course the function $\omega \mapsto | x |_\omega$ is integrable on (Ω, Σ, μ) for each $x \in c_{00}$, and so we can consider the seminorm $\| \cdot \|_0$ on c_{00} as defined in the preceding theorem. Moreover, the canonical unit vector sequence (e_i) in c_{00} is a c_0-sequence for $\| \cdot \|_0$ because

$$\| \sum_{i=1}^n a_i e_i \|_0 = \int_\Omega | \sum_{i=1}^n a_i e_i |_\omega \, d\mu(\omega) = \int_\Omega \| \sum_{i=1}^n a_i f_i(\omega) \| \, d\mu(\omega)$$

$$= \| \sum_{i=1}^n a_i f_i \|_1$$

Therefore, the preceding theorem implies that the set of all points $\omega \in \Omega$ for which (e_n) has a c_0-subsequence for $| \cdot |_\omega$ is a non null set. To complete the proof it is enough to observe that "(e_n) has a c_0-subsequence for $| \cdot |_\omega$" simply means that $(f_n(\omega))$ has a c_0-subsequence.

Theorem 2.1.3. *Let (Ω, Σ, μ) be a finite measure space and let X be a Banach space, then $L_1(\mu, X)$ contains a copy of c_0 if and only if X does.*

Proof. The condition is of course sufficient. Necessity follows from the preceding theorem.

Our aim now is to show that the preceding results are also true for $L_p(\mu, X)$ with $1 < p < +\infty$. To prove this we will use the following "subsequence splitting lemma" for $L_1(\mu)$-bounded sequences. It was discovered by Rosenthal [113], as a refinement of results and arguments due to Kadec and Pełczyński [77]. As far as we know it is the sharpest result of its kind.

Lemma 2.1.3 (Kadec-Pełczyński-Rosenthal, [77, 113]). *Let (Ω, Σ, μ) be a finite measure space. If (f_n) is a bounded sequence in $L_1(\mu)$, then there exist a subsequence (f_{n_k}) of (f_n) and a sequence (A_k) of pairwise disjoint measurable sets such that $(\chi_{\Omega \setminus A_k} f_{n_k})$ is uniformly integrable.*

Proof. Of course if (f_n) is uniformly integrable the result is trivial: it is enough to take $(f_{n_k}) = (f_n)$ and $A_n = \emptyset$ for all $n \in \mathbb{N}$. So, let us suppose that (f_n) is not uniformly integrable. This means that there exists $\epsilon > 0$ satisfying there exist a sequence (E_j) of measurable sets with $\lim_j \mu(E_j) = 0$, and a subsequence (f_{n_j}) of (f_n) such that

(*)
$$\int_{E_j} |f_{n_j}| \, d\mu > \epsilon$$

for all $j \in \mathbb{N}$

Let ϵ_0 be the supremum of all ϵ's for which (*) holds. It is clear that we can take a subsequence of (f_n), which we continue to denote in the same way, and a sequence (E_n) of measurable sets such that

$$\mu(E_n) < \frac{1}{2^n} \quad \text{and} \quad \int_{E_n} |f_n| \, d\mu > \epsilon_0 - \frac{1}{2^n}$$

for all $n \in \mathbb{N}$. Take $n_1 = 1$. Since

$$\lim_k \mu(\cup_{n=k}^\infty E_n) = 0,$$

by the absolute continuity of the indefinite integral of $|f_{n_1}|$, there exists $n_2 > n_1$ such that

$$\int_{E_{n_1} \setminus \cup_{n=n_2}^\infty E_n} |f_{n_1}| \, d\mu > \epsilon_0 - \frac{1}{2^{n_1}}.$$

Put

$$A_1 = E_{n_1} \setminus \cup_{n=n_2}^\infty E_n.$$

Now, applying the same argument to $|f_{n_2}|$, we get $n_3 > n_2$ such that

$$\int_{E_{n_2} \setminus \cup_{n=n_3}^\infty E_n} |f_{n_2}| \, d\mu > \epsilon_0 - \frac{1}{2^{n_2}}.$$

Put

$$A_2 = E_{n_2} \setminus \cup_{n=n_3}^\infty E_n.$$

It is clear that in this way we obtain a sequence (A_k) of pairwise disjoint measurable sets, and a subsequence (f_{n_k}) of (f_n) such that

$$\int_{A_k} |f_{n_k}| \, d\mu > \epsilon_0 - \frac{1}{2^{n_k}}$$

for all $k \in \mathbb{N}$. Let us finish the proof showing that $(\chi_{\Omega \setminus A_k} f_{n_k})$ is uniformly integrable. Suppose it is not the case. Then there exist $\epsilon' > 0$, a subsequence of (f_{n_k}), which we continue to denote in the same way, and a sequence (B_k) of measurable sets with $\lim_k \mu(B_k) = 0$, such that

$$\int_{B_k} |\chi_{\Omega \setminus A_k} f_{n_k}| \, d\mu > \epsilon'$$

for all $k \in \mathbb{N}$. But this implies that

$$\int_{B_k \cup A_k} |f_{n_k}| \, d\mu = \int_{A_k} |f_{n_k}| \, d\mu + \int_{B_k} \chi_{\Omega \setminus A_k} |f_{n_k}| \, d\mu > \epsilon_0 - \frac{1}{2^{n_k}} + \epsilon'$$

for all $k \in \mathbb{N}$. Since $\epsilon_0 - \frac{1}{2^{n_k}} + \epsilon' > \epsilon_0$ for k large enough, this is a contradiction with the fact that ϵ_0 is the supremum of the ϵ's satisfying (*).

We can now begin our study of $L_p(\mu, X)$ spaces. Next theorem is a good example of how the preceding lemma can be used to deal with vector-valued functions.

Theorem 2.1.4. *Let (Ω, Σ, μ) be a finite measure space, let $1 < p < +\infty$, and let $i : L_p(\mu, X) \hookrightarrow L_1(\mu, X)$ be the canonical embedding. Let H be a subspace of $L_p(\mu, X)$ which does not contain a complemented copy of ℓ_p, then the restriction $i|_H$ of i to H is an isomorphic embedding.*

Proof. If $i|_H$ is not an isomorphic embedding, then not all $\| \cdot \|_1$-null sequences in H are $\| \cdot \|_p$-null. Thus, using the continuity of i, it follows that there exists a $\| \cdot \|_1$-null sequence (f_n) in H such that

$$\| f_n \|_p = 1$$

for all $n \in \mathbb{N}$. Taking a subsequence if necessary, we may suppose that $(f_n(\omega))$, and therefore $(\| f_n(\omega) \|^p)$, goes to zero for almost all $\omega \in \Omega$. Now we can apply the preceding lemma to the sequence $(\| f_n(\cdot) \|^p)$, and we get a further subsequence, which we still denote (f_n), and a sequence (A_n) of pairwise disjoint measurable sets such that $(\chi_{\Omega \setminus A_n} \| f_n(\cdot) \|^p)$ is uniformly integrable. An appeal to Vitali's lemma [51, IV.10.9.] allows us to conclude that $(\chi_{\Omega \setminus A_n} \| f_n(\cdot) \|^p)$ is null in $L_1(\mu)$, that is, $(\chi_{\Omega \setminus A_n} f_n)$ is null in $L_p(\mu, X)$. Therefore, taking again a subsequence if necessary, we have that $(\chi_{A_n} f_n) = (f_n - \chi_{\Omega \setminus A_n} f_n)$ is a seminormalized sequence in $L_p(\mu, X)$. Since the functions are disjointly supported, $(\chi_{A_n} f_n)$ is a complemented ℓ_p-sequence in $L_p(\mu, X)$ (Proposition 1.4.1). At this point, the classical result on perturbation of basis (see [93, 1.a.9] or [35, Chapter V, Theorem 12]), guarantees that

$$(f_n) = (\chi_{A_n} f_n + \chi_{\Omega \setminus A_n} f_n)$$

has an ℓ_p-subsequence which is complemented in $L_p(\mu, X)$, and therefore in H. This completes the proof.

Remark 2.1.1. Notice that it is possible to give more precise versions of the preceding theorem. For instance, it is also true for the canonical injection from $L_p(\mu, X)$ into $L_r(\mu, X)$, with $p > r \geq 1$. We may also deduce from the proof that it also holds for all subspaces H which do not contain sequences arbitrarily close to normalized disjointly supported sequences, and in particular, which does not contain $(1 + \epsilon)$-complemented $(1 + \epsilon)$-ℓ_p-sequences for arbitrarily small $\epsilon > 0$.

As an immediate consequence of the preceding theorem we have:

Corollary 2.1.1. *Let (Ω, Σ, μ) be a finite measure space and let $1 < p < +\infty$. Then we have:*

(a) Every c_0-sequence in $L_p(\mu, X)$ is a c_0-sequence in $L_1(\mu, X)$.

(b) Every ℓ_1-sequence in $L_p(\mu, X)$ is an ℓ_1-sequence in $L_1(\mu, X)$.

(c) Every ℓ_r-sequence in $L_p(\mu, X)$ is an ℓ_r-sequence in $L_1(\mu, X)$, whenever $1 \leq r < +\infty$, $r \neq p$.

(d) If $L_p(\mu, X)$ contains a copy of H and H does not contain complemented copies of ℓ_p, then $L_1(\mu, X)$ contains a copy of H.

We can now extend Theorem 2.1.2 to $L_p(\mu, X)$ spaces with $1 \leq p < +\infty$. In the extension we also remove the finiteness assumption on μ.

Theorem 2.1.5. *Let (Ω, Σ, μ) be an arbitrary measure space, let $1 \leq p < +\infty$, and let (f_n) be a c_0-sequence in $L_p(\mu, X)$. Then, the set of all $\omega \in \Omega$ such that $(f_n(\omega))$ has a c_0-subsequence is a measurable set with positive measure.*

Proof. Let (f_n) be a c_0-sequence in $L_p(\mu, X)$. Notice first that if μ is a finite measure, we can apply the preceding corollary, and we deduce that (f_n) is a c_0-sequence in $L_1(\mu, X)$, too. Now, the conclusion follows from Theorem 2.1.2.

For the general case, we use the standard reduction argument we have already given in Corollary 1.6.1. By Lemma 1.6.1, there exists a σ-finite measure space $(\Omega_1, \Sigma_1, \mu_1) \subset (\Omega, \Sigma, \mu)$ such that (f_n) lies in $L_p(\mu_1, X)$. Now, by Proposition 1.6.1, there exists a probability measure μ_0 on (Ω_1, Σ_1) and an isometric isomorphism from $L_p(\mu_1, X)$ onto $L_p(\mu_0, X)$. Moreover, as we showed in the proof, this isometric isomorphism has the form

$$f \longrightarrow hf$$

where h is a certain scalar function which does not vanish at any point. Therefore, it is clear that $(h(\omega)f_n(\omega))$ has a c_0-subsequence if an only if so does $(f_n(\omega))$. So the conclusion follows from the first part.

Finally, we get the following

Theorem 2.1.6 (Kwapień). *Let (Ω, Σ, μ) be an arbitrary measure space and let $1 \leq p < +\infty$, then $L_p(\mu, X)$ contains a copy of c_0 if and only if X contains a copy of c_0.*

2.2 When does $L_p(\mu, X)$ contain a copy of ℓ_1?

Our initial point of view here is very similar to the one in the preceding section. In this way we may see that the natural modification of Example 2.1.1 shows that for $1 < p < +\infty$ there are ℓ_1-sequences (f_n) in $L_p([0,1], \ell_1)$ such that for each subsequence (f_{n_k}) of (f_n), $(f_{n_k}(t))$ is not an ℓ_1-sequence for (almost) any $t \in [0,1]$. So, likewise in the preceding section, it is natural to ask whether given an ℓ_1-sequence (f_n) in $L_p(\mu, X)$, we can assure that the set of all $\omega \in \Omega$ such that $(f_n(\omega))$ has an ℓ_1-*subsequence* is a non null set. The main theorem of the section states that we can.

We begin our work with quite a general result about sequences in $L_p([0,1], X)$ due to Batt and Hiermeyer. Recall that (r_n) denotes the sequence of Rademacher functions in $[0,1]$.

Proposition 2.2.1 (Proposition 4.1 of [3]). *Let (x_n) be a bounded sequence in X and let $1 \le p < +\infty$. Then, the following are equivalent:*

(a) (x_n) has no ℓ_1-subsequences.
(b) $(r_n(.)x_n)$ is a weakly null sequence in $L_p([0,1], X)$.

Proof. Suppose that (a) holds. In order to conclude (b) we will see that every subsequence of $(r_n(.)x_n)$ has a further subsequence which is weakly null in $L_p([0,1], X)$. But Rosenthal's theorem says us that each subsequence of (x_n) has a weakly Cauchy subsequence. Thus, we only need to show

(*) If (y_k) is a weakly Cauchy sequence in X and (r_{n_k}) is a subsequence of (r_n), then $(r_{n_k}(.)y_k)$ is weakly null in $L_p([0,1], X)$.

So, assume we are in the hypothesis of (*) and let Γ be a functional in $L_p([0,1], X)^*$. By theorem 1.5.4, there exists a w^*-measurable function $\Psi : [0,1] \to X^*$, such that $\| \Psi(.) \|$ is measurable, it belongs to $\mathcal{L}_q([0,1])$ (as usual, q denotes the conjugate of p) and

$$\Gamma(f) = \int_0^1 < f(t), \Psi(t) > dt$$

for all $f \in L_p([0,1], X)$. The sequence $(< y_k, \Psi(t) >)_k$ converges for almost all $t \in [0,1]$ because (y_k) is weakly Cauchy. Let us denote by h the pointwise limit of $(< y_k, \Psi(.) >)_k$. Since

$$|< y_k, \Psi(t) >| \le sup_n \|y_n\| \, \|\Psi(t)\|$$

for all $t \in [0,1]$ and all $k \in \mathbb{N}$, we can use Lebesgue's dominated convergence theorem. Therefore, h belongs to $L_q([0,1])$ and $(< y_k, \Psi(.) >)_k$ converges to h in $L_q([0,1])$. Now , note that

$$
\begin{aligned}
| \Gamma(r_{n_k}(.)y_k) | &= \left| \int_0^1 < r_{n_k}(t)y_k, \Psi(t) > dt \right| \\
&= \left| \int_0^1 r_{n_k}(t) < y_k, \Psi(t) > dt \right| \\
&\le \left| \int_0^1 r_{n_k}(t) \left(< y_k, \Psi(t) > - h(t) \right) dt \right| + \left| \int_0^1 r_{n_k}(t) h(t) \, dt \right|
\end{aligned}
$$

$$\leq \left(\int_0^1 | <y_k, \Psi(t)> -h(t)|^q \, dt \right)^{\frac{1}{q}} + \left| \int_0^1 r_{n_k}(t)h(t) \, dt \right|$$

Using the fact that (r_{n_k}) is weakly null in $L_p([0,1])$, it is clear that $(\Gamma(r_{n_k}(.)y_k))$ is a null sequence. Thus we have shown that (a) implies (b). For the converse, it is enough to observe that, for $1 \leq p < +\infty$, $(r_n(.)x_n)$ is an ℓ_1-sequence in $L_p([0,1], X)$ whenever (x_n) is an ℓ_1-sequence in X: if $\delta > 0$ satisfies

$$\| \sum \lambda_n x_n \| \geq \delta \sum | \lambda_n |$$

for all finite sequences (λ_n) of scalars, then we have

$$\| \sum \lambda_n r_n(.)x_n \|_p = \left(\int_0^1 \| \sum \lambda_n r_n(t)x_n \|^p \, dt \right)^{1/p}$$

$$\geq \left(\int_0^1 \delta^p \sum | \lambda_n r_n(t) |^p \, dt \right)^{1/p} = \delta \sum | \lambda_n |$$

for every finite sequence (λ_n) of scalars. This finishes the proof.

Our next lemma provides a very easy sufficient condition for a sequence in $L_1(\mu, X)$ to be weakly null.

Lemma 2.2.1. *Let (Ω, Σ, μ) be a finite measure space, let (f_n) be a bounded sequence in $L_\infty(\mu, X)$, and let us suppose that $(f_n(\omega))$ is weakly null for almost all $\omega \in \Omega$. Then (f_n) is weakly null in $L_1(\mu, X)$.*

Proof. By Theorem 1.5.4 we only need to show that if we take a w^*-measurable function $\Psi : \Omega \to X^*$ such that the function $\| \Psi(.) \|$ is measurable and belongs to $\mathcal{L}_\infty(\mu)$, then

$$\int_\Omega < f_n(\omega), \Psi(\omega) > d\mu \to 0.$$

But this clearly follows from the Lebesgue's dominated convergence theorem, because $\lim_n < f_n(\omega), \Psi(\omega) > = 0$ for almost all $\omega \in \Omega$ and

$$|< f_n(\omega), \Psi(\omega) >| \leq \| f_n(\omega) \| \, \| \Psi(\omega) \| \leq \sup_n \| f_n(\omega) \| \, \| \Psi(\omega) \| .$$

We can now give the main result of the section.

Theorem 2.2.1 (Maurey-Pisier-Bourgain, 1978). *Let (Ω, Σ, μ) be an arbitrary measure space, let $1 \leq p < +\infty$, and let (f_n) be an ℓ_1-sequence in $L_p(\mu, X)$. Let us assume that one of the following two conditions holds*

(a) μ is finite, $p = 1$ and (f_n) is uniformly integrable.
(b) $1 < p < +\infty$.

Then, there the set of all $\omega \in \Omega$ such that $(f_n(\omega))$ has an ℓ_1-subsequence is a measurable set with positive measure.

Proof. By Corollary 1.6.1 we only have to show that the set of all $\omega \in \Omega$ such that $(f_n(\omega))$ has an ℓ_1-subsequence is non null.

First, let us consider case (a). By the hypothesis, there exist two positive numbers M, δ such that

$$\delta \sum |a_n| \leq \| \sum a_n f_n \|_1 \leq M \sum |a_n|$$

for all finite sequences (a_n) of scalars. Since $(\| f_n(.) \|)$ is uniformly integrable, by standard arguments (see for instance [35, page 111]), it follows that there exist $N > 0$ and a sequence (A_n) of measurable sets such that for every $n \in \mathbb{N}$ we have

$$\int_{\Omega \setminus A_n} \| f_n(\omega) \| \, d\mu(\omega) < \frac{\delta}{2} \quad \text{and} \quad \| f_n(\omega) \| \leq N \ \text{ for all } \ \omega \in A_n.$$

Take $g_n = \chi_{A_n} f_n$ for every $n \in \mathbb{N}$. Notice that (g_n) is bounded in $L_1(\mu, X)$ (even in $L_\infty(\mu, X)$) and for every finite sequence (a_n) of scalars we have

$$\| \sum a_n g_n \|_1 \ \geq \ \| \sum a_n f_n \|_1 - \| \sum a_n (f_n - g_n) \|_1$$

$$\geq \ \delta \sum |a_n| - \frac{\delta}{2} \sum |a_n| = \frac{\delta}{2} \sum |a_n|$$

Hence, (g_n) is an ℓ_1-sequence. Let us proceed now by contradiction. If $(f_n(\omega))$ has no ℓ_1-subsequences for almost all $\omega \in \Omega$, then the same holds for the sequence (g_n). But $(g_n(\omega))_n$ is bounded in X for almost all $\omega \in \Omega$, and so, according to Proposition 2.2.1, $(r_n(.)g_n(\omega))$ is a weakly null sequence in $L_1([0,1], X)$ for almost all $\omega \in \Omega$. Let us denote by $r_n g_n(.)$ the function

$$\begin{aligned} \Omega &\longrightarrow L_1([0,1], X) \\ \omega &\longrightarrow r_n g_n(\omega) \end{aligned}$$

Notice that the sequence $(r_n g_n(.))$ is bounded in $L_\infty(\mu, L_1([0,1], X))$ because

$$\| r_n g_n(.) \|_\infty \ = \ \text{ess sup}_{\omega \in \Omega} \| r_n g_n(\omega) \|_1 = \text{ess sup}_{\omega \in \Omega} \int_0^1 \| r_n(t) g_n(\omega) \| \, dt$$

$$= \ \text{ess sup}_{\omega \in \Omega} \| g_n(\omega) \| \leq N$$

Thus, by the preceding lemma, it follows that the sequence $(r_n g_n(.))$ is weakly null in $L_1(\mu, L_1([0,1], X))$. Since this space is isometrically isomorphic in the natural way to $L_1([0,1], L_1(\mu, X))$, we deduce that $(r_n(.)g_n)$ is weakly null in this last space, where of course $(r_n(.)g_n)$ denote the functions

$$\begin{aligned} [0,1] &\longrightarrow L_1(\mu, X) \\ t &\longrightarrow r_n(t) g_n \end{aligned}$$

Again, an appeal to Proposition 2.2.1 says us that (g_n) has no ℓ_1-subsequences in $L_1(\mu, X)$. This contradiction finishes the proof in the case $p = 1$.

For the case (b), assume first the measure μ is finite. In this situation, notice that (f_n) is uniformly integrable, just because it is bounded in $L_p(\mu, X)$ with $p > 1$. On the other hand, it follows from Corollary 2.1.1 that (f_n) is an ℓ_1-sequence in $L_1(\mu, X)$. Therefore, our sequence is actually in the hypothesis of case (a). So the conclusion holds. For an arbitrary measure we just have to follow the same procedure we used in Theorem 2.1.5.

As an immediate consequence of this theorem we get

Theorem 2.2.2 (Pisier [101], 1978). *Let (Ω, Σ, μ) be an arbitrary measure space and let $1 < p < +\infty$, then $L_p(\mu, X)$ contains a copy of ℓ_1 if and only if X does.*

Remark 2.2.1. One could think that the finiteness assumption on μ in part (a) of Theorem 2.2.1 could be removed. However, this is not the case as the following simple example shows:

Take any uniformly bounded sequence (h_n) of norm one non negative functions in $L_1([0, +\infty))$ such that

$$h_n(t) = 0 \quad \text{for all} \ \ t \in [0, n]$$

for all $n \in \mathbb{N}$. Let (e_n) be the canonical ℓ_1-basis. It is not difficult to to show that if we take

$$f_n = h_n(.)e_n,$$

then (f_n) is an uniformly integrable ℓ_1-sequence in $L_1([0, +\infty), \ell_1)$. However, for each $t \in [0, +\infty)$ the sequence $(f_n(t))$ is eventually null and so, it can not have any ℓ_1-subsequence.

2.3 Notes and Remarks

The importance of Hoffmann-Jørgensen's paper [73] in Kwapień's theorem 2.1.6, the main result of section 2.1, can not be overestimated. Kwapień's short note [83] was published in the same issue in which the aforementioned paper appeared, and even its title ("On Banach spaces containing c_0. A supplement to the paper by J. Hoffmann-Jørgensen 'Sums of independent Banach space valued random variables' ") reveals how it depends on it.

Hoffmann-Jørgensen's fundamental paper [73] is one of the pioneering works on probability in Banach spaces. It is focused on generalizing several classical theorems about sums of independent *real* random variables to the case of *Banach space valued* random variables. Among other results, it is showed that for a Banach space X the following are equivalent:

(a) X satisfies: "If (φ_n) is a sequence of independent, symmetric, X-valued random variables and their partial sums $(S_m) = (\sum_{n=1}^m \varphi_n)$ are bounded almost everywhere, then (S_m) converges almost everywhere".

(b) $L_p([0,1], X)$ does not contain a copy of c_0 for *some* $p \in [1, +\infty)$.

(c) $L_p([0,1], X)$ does not contain a copy of c_0 for *any* $p \in [1, +\infty)$.

These equivalences led Hoffmann-Jørgensen to conjecture that the preceding conditions would also be equivalent to the following one

(d) X does not contain a copy of c_0.

Kwapień [83] showed that the conjecture was true.

In 1978, four years after the publication of the papers of Hoffmann-Jørgensen and of Kwapień, Bourgain [14] provided an alternative proof of Kwapień's theorem obtaining a more general result. Actually, we must say that Bourgain's results are formulated in a more general setting than required now, but, as we have already remarked, later on his solution will be crucial for us (see Saab and Saab's theorem 3.1.4). This is why we present here Bourgain's approach.

We do not use the probability language and no particular knowledge of probability is assumed. However, its flavour is all over Section 2.1. This should not be surprising because, as we have explained, the results considered here had their roots precisely in one of the pioneering works on probability in Banach spaces.

We knew about Kadec-Pełczyński-Rosenthal's lemma 2.1.3 through an unpublished paper of Bourgain [17], and we learnt there how to apply it in the study of vector-valued functions. Although it is mentioned by passing in some places in the literature (see for instance [19]), we are afraid that it is widely ignored. However, we believe that it is a very important result which deserves a lot of attention. We have known through Bourgain and Rosenthal that this "subsequence splitting lemma" was formulated for the first time by the second named author in the spring of 1979, in the *Topics Course at the University of Paris VI*.

Theorem 2.1.4 is perhaps well known for many specialist but, as far as we know, it appears explicitly here for the first time. The same can be said about the case $p > 1$ in Theorem 2.1.5. Some of the particular cases of Theorem 2.1.4, which we give in its Corollary, may be found included in some proofs in the literature (see, for instance, [108], [69] or [10]).

Also in 1978, Pisier, answering a question of Diestel, showed that, for $1 < p < +\infty$, $L_p(\mu, X)$ contains a copy of ℓ_1 if and only if X does. He remarked in his paper: *"soon after my proof, B. Maurey gave another proof of the result giving the following improvement* (he is referring to theorem 2.2.1); *I could verify that my proof gave the same extension, too"*. Since Bourgain also asserts in [16] that he got independently the same result, we have called it Maurey-Pisier-Bourgain theorem. Bourgain's proof is quite complicated, via an averaging result for ℓ_1-sequences. Our proof follows very much Pisier's approach.

We must say that Batt and Hiermeyer's proposition 2.2.1 has a clear antecedent in Proposition 1 of [101], which in turn, in Pisier's own words, is *"just a reformulation of a remark of Stegall and Odell"* [112, Addendum, p. 377]. It is one of the main tools used by Pisier to prove his theorem.

In our formulations of Bourgain's Theorem 2.1.5 and Maurey-Pisier-Bourgain's Theorem 2.2.1, we have taken advantage of the measurability results obtained in Section 1.6. In fact, in the original formulations of these theorems, no assertion about the measurability of the the set of all points $\omega \in \Omega$ for which $(f_n(\omega))$ has either a c_0-subsequence or an ℓ_1-subsequence was made. We thought natural to study this, and so we were led to Proposition 1.6.3

All Section 2.2 is clearly connected with the problems of weak compactness, weak convergence, and ℓ_1-sequences in $L_p(\mu, X)$ spaces. We have to point out that Talagrand's fundamental paper [127] must be studied by anyone interested in such problems. In particular in [127] one can find much stronger results than our simple lemma 2.2.1. Besides Talagrand's paper, the following papers should also be mentioned: [129], [38], [9], [30] and [33].

Once we have touched on weak compactness in $L_p(\mu, X)$ spaces one can not help wondering about *strong* compactness in these spaces. In this line we have some quite recent papers: [62, 63], [2] and [32].

Finally, we would like to call the attention here to an important line of research which has some connection with the problems studied in this Chapter: vector-valued versions of results on basis properties of $L_p(\mu)$ spaces. Recall that for $1 < p < +\infty$ and μ finite:

(a) $L_p(\mu)$ has an unconditional basis.
(b) Every sequence of non zero elements in $L_p(\mu)$ which is a martingale difference sequence is an unconditional basic sequence.

Diestel and Uhl [39] asked if $L_p(\mu, X)$ has an unconditional basis whenever X has, and in general they asked for vectorial versions of these results. The first question was soon negatively answered by Aldous [1], who found a necessary condition for $L_p(\mu, X)$ to have an unconditional basis. We believe that at the moment it is unknown whether it is sufficient or not. Concerning the second question, we can say that the Banach spaces for which assertion (b) is true have been extensively studied (see [21, 22, 23]). They are the so-called UMD spaces, which turned out to be the natural context in many problems in analysis.

3. $C(K, X)$ spaces

After the fundamental results on $L_p(\mu, X)$ spaces of the preceding chapter, let us see now which is the behavior of $C(K, X)$ spaces. We will study the contributions of E. and P. Saab, P. Cembranos and F. Freniche, and L. Drewnowski. They provide a complete answer to the problems we are dealing with.

The questions: "does $L_p(\mu)$ have a copy of c_0, ℓ_1 or ℓ_∞?", "does $L_p(\mu)$ have a complemented copy of c_0 or ℓ_1?" have quite an easy answer, and it does not depend on the measure space (Ω, Σ, μ) (remember we are always assuming $L_p(\mu)$ infinite-dimensional). The situation is completely different for $C(K)$ spaces. We know that $C(K)$ *always* contains a copy of c_0 (see Section 1.4), and we also know that it *never* contains a *complemented* copy of ℓ_1 (this is not difficult and it will be shown in the next section), but if we wish to know whether $C(K)$ contains a copy of ℓ_1, or a copy of ℓ_∞, or a complemented copy of c_0, then the answer will be: "It depends on the compact space K. For some of them the answer is yes, and for some of them the answer is no". Of course, we would like to know for which compact spaces the answer is "yes", that is, we would like to have a charaterization of these facts through topological properties of K. In this line we have a satisfactory answer for copies of ℓ_1 (Theorem 3.1.1), but for complemented copies of c_0 and for copies of ℓ_∞, the problems are still open.

3.1 Copies and complemented copies of ℓ_1 in $C(K, X)$

Let us begin characterizing when $C(K, X)$ contains a copy of ℓ_1. We should first know the situation in the scalar case. For this reason we have to recall quite a classical and well known result due to Pełczyński and Semadeni [100]. To understand it we need to remember that a compact Hausdorff space is said to be *scattered* (or *dispersed*) if each of its closed subsets have isolated points (see [84] or [124]). The simplest example of a scattered compact space is the Alexandroff compactification of the natural numbers, and in general any countable and compact space. A typical example of a non scattered compact space is $[0, 1]$.

The following theorem is just [84, Section 13, Corollary to Theorem 4] (see also [124, Section 8.5]).

Theorem 3.1.1 (Pełczyński-Semadeni [100], 1959). *Let K be a compact and Hausdorff space, then $C(K)$ contains a copy of ℓ_1 if and only if K is not scattered.*

Now we can study the vector-valued case. The following result is perhaps well known and very old, but we have only found it in [25].

Theorem 3.1.2. *Let K be a compact and Hausdorff space, then $C(K, X)$ contains a copy of ℓ_1 if and only if either X contains a copy of ℓ_1 or K is not scattered (equivalently, $C(K)$ contains a copy of ℓ_1).*

Proof. Let us show the non trivial implication. Assume K scattered and let us suppose that $C(K, X)$ contains a copy of ℓ_1. Since a Hausdorff space which is the range of a continuous map defined on a compact scattered space is also a compact scattered space [124, II.8.5.3.], an old technique (see for example the proof of [51, VI.7.6.]) allows us to assume also that K is metrizable. But it is well known that a scattered metrizable compact space must be countable (see [124, 8.5.5.] or [84, Section 5, Corollary 2]), so K must be countable. Then, put $K = \{t_m : m \in \mathbb{N}\}$ and let (ϕ_n) be an ℓ_1-sequence in $C(K, X)$. If X does not contain ℓ_1 then Rosenthal's theorem says that, for each $m \in \mathbb{N}$, $(\phi_n(t_m))_n$ has a weakly Cauchy subsequence. But then, by a standard diagonal argument, we can deduce that (ϕ_n) has a subsequence, which we continue to denote in the same way, such that $(\phi_n(t_m))_n$ is weakly Cauchy for all $m \in \mathbb{N}$. Therefore, by Corollary 1.7.1, (ϕ_n) is weakly Cauchy in $C(K, X)$. Of course, this contradicts the fact that it is an ℓ_1-sequence. $\blacksquare$

The problem of characterizing when $C(K, X)$ contains a *complemented* copy of ℓ_1 was solved by E. and P. Saab [116].

To understand E. and P. Saab's proof let us begin thinking how to prove that $C(K)$ does not contain a *complemented* copy of ℓ_1 (remember that many $C(K)$ spaces *do contain* copies of ℓ_1, for instance $C([0,1])$, see also Theorem 3.1.1 above). Perhaps the simplest proof of this is the following:

Let us suppose that $C(K)$ contains a complemented copy of ℓ_1. Then $C(K)^*$, the dual of $C(K)$, which we identify with the space $rcabv(K)$ of all regular measures of bounded variation on the Borel subsets of K (see Section 1.7), contains a copy of ℓ_∞, and therefore it contains a c_0-sequence (μ_n). Take

$$\lambda = \sum_{k=1}^{\infty} \frac{1}{2^k} \, \|\mu_k\|,$$

where $\|\mu_k\|$ denotes the variation of μ_k. Radon-Nikodym theorem allows us to consider (μ_n) as a sequence in $L_1(\lambda)$ and so we would have a copy of c_0 in $L_1(\lambda)$. Of course this is a contradiction (use for example that $L_1(\lambda)$ is weakly sequentially complete).

Observe now that the preceding proof generalizes easily to prove that if $C(K,X)$ contains a complemented copy of ℓ_1 then X must contain a complemented copy of ℓ_1, *for spaces X whose dual X^* enjoys Radon-Nikodym property*. It is enough to apply Kwapień 's Theorem 2.1.3 and the classical Bessaga-Pełczyński theorem asserting that a Banach space X contains a complemented copy of ℓ_1 if and only if X^* contains a copy of c_0 (Theorem 10, Chapter V of [35]). E. and P. Saab realized that it was possible to overcome the restriction on X following carefully the representation of the dual of $L_p(\mu,X)$ provided by lifting theory (see Section 1.5), and using Bourgain's Theorem 2.1.1 instead of Kwapień 's Theorem 2.1.3. We believe that these are the main points in Saab & Saab's approach. We have summarized it in the following result, which is not explicitly stated in [116], but it is clearly contained in the proof given there.

At first, one might be surprised by the statement of the next theorem just because it seems it does not have much to do with $C(K,X)$. However, notice that the dual of $C(K,X)$, which we identify with $rcabv(K,X^*)$, is just a subspace of $cabv(\mathcal{B}(K),X^*)$, a particular case of the space involved in the statement. On the other hand, we must say that the full generality of our statement will be used in Section 4.1 (Theorem 4.1.1).

Theorem 3.1.3 (E. and P. Saab [116], 1982). *Let X be a Banach space, then $cabv(\Sigma,X^*)$ contains a copy of c_0 if and only if X^* does.*

Proof. Assume $cabv(\Sigma,X^*)$ contains a copy of c_0. Let (m_n) be a c_0-sequence in $cabv(\Sigma,X^*)$. Then there are $C_1, C_2 > 0$ such that

$$C_1 \max_i | a_i | \leq \| \sum_i a_i m_i \| \leq C_2 \max_i | a_i |$$

for all $(a_i) \in c_{00}$. Take $\lambda = \sum_{n=1}^{\infty} \frac{1}{2^n} \| m_n \|$. It is clear that (m_n) is a sequence in $cabv_\lambda(\Sigma,X^*)$, and we can assume λ to be complete (if necessary we λ-complete de σ-field Σ and we extend by zero the m_n's). Take a lifting ρ in $L_\infty(\lambda)$, and use Theorem 1.5.2. We get a sequence $(\varphi_n) = (\hat{\rho}(m_n))$ in $\mathcal{L}^{\infty}_{w^*}(\lambda,X^*)$ satisfying:

$$C_1 \max_i | a_i | \leq \| \sum_i a_i m_i \| = \int_\Omega \| \hat{\rho}(\sum_i a_i m_i)(\omega) \| \, d\lambda(\omega)$$

$$= \int_\Omega \| \sum_i a_i \varphi_i(\omega) \| \, d\lambda(\omega) \leq C_2 \max_i | a_i |$$

for all $(a_i) \in c_{00}$. Then we deduce from Bourgain's Theorem 2.1.1 that X^* must contain a copy of c_0 (recall the proof of Theorem 2.1.2).

An immediate consequence of the preceding result is the following

Theorem 3.1.4 (E. and P. Saab [116], 1982). $C(K,X)$ *contains a complemented copy of ℓ_1 if and only if X does.*

Proof. If $C(K, X)$ contains a complemented copy of ℓ_1 then $rcabv(K, X^*)$, the dual of $C(K, X)$, contains a copy of ℓ_∞, and therefore, a copy of c_0. Of course, this implies that $cabv(\mathcal{B}(K), X^*)$ ($\supset rcabv(K, X^*)$) also contains a copy of c_0. The preceding theorem implies that X^* contains a copy of c_0. But, by the classical result of Bessaga and Pełczyński mentioned above (Theorem 10, Chapter V of [35]), this means that X contains a complemented copy of ℓ_1.

3.2 Complemented copies of c_0 in $C(K, X)$

We have seen that $C(K)$ (and therefore $C(K, X)$) always contains many copies of c_0 (see Section 1.4). However, what can be said about *complemented* copies of c_0? Notice first that in general $C(K)$ *does not* contain *complemented* copies of c_0. For example take ℓ_∞, which is just $C(\beta\mathbb{N})$, where as usual, $\beta\mathbb{N}$ denotes the Stone-Čech compactification of the natural numbers. Phillips classical theorem (Corollary 1.3.2) guarantees that $C(\beta\mathbb{N})$ does not contain complemented copies of c_0. In general, to get a $C(K)$ space with no complemented copies of c_0, it is enough to take any extremally disconnected compact space K, because the corresponding $C(K)$ is a Grothendieck space (see [35] and the Notes and Remarks of this Chapter). On the other hand, we may find lots of Banach spaces not containing c_0 at all. However, we will see in the next theorem that $C(K, X)$ *always* contains a complemented copy of c_0 (remember that we are excluding the trivial cases of finite dimensional spaces $C(K)$ or X). Of course, as we have already mentioned, this was quite surprising and provided the first negative answer to Problem 2 in the Introduction. The result was obtained independently by Cembranos and Freniche. It is worth to mention that the main ingredient in its proof is Josefson-Nissenzweig Theorem [35, Chapter XII]. We have found the proof we give in Bombal's paper [5].

Theorem 3.2.1 (Cembranos-Freniche [26] [59], 1984). *The Banach space $C(K, X)$ always contains a complemented copy of c_0.*

Proof. By Josefson-Nissenzweig theorem, there exists a normalized weak*-null sequence (x_n^*) in X^*. Take a bounded sequence (x_n) in X such that $x_n^*(x_n) = 1$, and with this sequence (x_n), let us repeat the construction of a c_0-sequence we did just before Proposition 1.4.2. We get a c_0-sequence (g_n) in $C(K, X)$ such that

$$\| g_n \| = \| g_n(t_n) \|$$

for all $n \in \mathbb{N}$, where (t_n) is a sequence in K such that

$$g_n(t_m) = \begin{cases} x_n & \text{if } n = m \\ 0 & \text{otherwise} \end{cases}$$

Since (x_n^*) is w^*-null, it converges uniformly to zero on the compact subsets of X, and so

$$C(K, X) \quad \longrightarrow \quad C(K, X)$$

$$\phi \quad \longrightarrow \quad \sum_{n=1}^{\infty} x_n^*(\phi(t_n))g_n$$

is a well defined continuous linear projection onto the closed linear span of the (g_n)'s. This completes the proof.

Remark 3.2.1. It should be pointed out that if we wish to use the language of Proposition 1.1.2 in the preceding proof, we just have to realize that $\delta_{t_n}(.)x_n^*$ is a w^*-null sequence in $C(K, X)^*$ satisfying

$$<g_n, \delta_{t_m}(.)x_m^*> = <g_n(t_m), x_m^*> = \delta_{nm} <x_n, x_m^*> = \delta_{nm}$$

for all $n, m \in \mathbb{N}$, where of course the g_n's, x_n's, t_m's and x_m^*'s are the ones of the preceding proof.

3.3 Copies of ℓ_∞ in $C(K, X)$

Finally, we wish to know when $C(K, X)$ contains a copy of ℓ_∞. The problem was solved by Drewnowski. We will need to apply very carefully all the power of the results of Section 1.3.

Theorem 3.3.1 (Drewnowski [46], 1990). $C(K, X)$ *contains a copy of* ℓ_∞ *if and only if either* $C(K)$ *or* X *does.*

Proof. The condition is of course sufficient. Let us show that it is also necessary. To do this let us observe first the following fact:

Given a compact Hausdorff space K *and a Banach space* X, *if* $T : \ell_\infty \to C(K, X)$ *is a bounded linear operator such that*

$$T((\zeta_n))(t) = \sum_{n=1}^{\infty} \zeta_n T(e_n)(t)$$

for all $t \in K$ *and all* $(\zeta_n) \in \ell_\infty$, *then*

$$\lim_n T(e_n) = 0,$$

where (e_n) *denotes the canonical unit vector sequence in* ℓ_∞.

This fact may be proved using Remark 1.3.1, but instead of this result we will use Orlicz-Pettis Theorem ([35, page 24]). Let $T : \ell_\infty \to C(K,X)$ be a bounded linear operator such that

$$T((\zeta_n))(t) = \sum_{n=1}^{\infty} \zeta_n T(e_n)(t)$$

for all $t \in K$ and all $(\zeta_n) \in \ell_\infty$. Then for each $(\zeta_n) \in \ell_\infty$, the bounded sequence $(\sum_{n=1}^{k} \zeta_n T(e_n))_k = (T(\sum_{n=1}^{k} \zeta_n e_n))_k$ is pointwise convergent to $T((\zeta_n))$, and by Proposition 1.7.1, it is weakly convergent, too. This means that the series $\sum \zeta_n T(e_n)$ is weakly convergent for all $(\zeta_n) \in \ell_\infty$. Therefore, by Orlicz-Pettis theorem ([35, page 24]), the series $\sum T(e_n)$ is unconditionally convergent. Of course this implies that $\lim_n T(e_n) = 0$.

Our aim now is to show that if $C(K,X)$ contains a copy of ℓ_∞ but neither $C(K)$ nor X does then it is possible to find a closed subset K_0 of K and a bounded linear operator $T : \ell_\infty \to C(K_0, X)$ such that

$$T((\zeta_n))(t) = \sum_{n=1}^{\infty} \zeta_n T(e_n)(t)$$

for all $t \in K_0$ and all $(\zeta_n) \in \ell_\infty$, but

$$\|T(e_n)\| \nrightarrow 0.$$

This is a contradiction with the fact we have just shown and our proof will be complete.

So, let us assume that $C(K,X)$ contains a copy of ℓ_∞ and suppose that neither $C(K)$ nor X contain a copy of ℓ_∞. Let $J : \ell_\infty \to C(K,X)$ be an isomorphic embedding, and let us denote $f_n = J(e_n)$. For each $t \in K$ consider the bounded linear operator

$$\begin{array}{rcl} J_t : \ell_\infty & \longrightarrow & X \\ (\zeta_n) & \longrightarrow & J((\zeta_n))(t) \end{array}$$

Since we are assuming that X does not contain ℓ_∞, it follows that all these operators are weakly compact and that $\sum J_t(e_n) = \sum f_n(t)$ is an unconditionally convergent series for all $t \in K$ (see Corollary 1.3.1 and remark 1.3.2), that is,

$$\sum \zeta_n f_n(t) \text{ is convergent} \tag{3.1}$$

for all $(\zeta_n) \in \ell_\infty$. Let D be a countable subset of K such that for each $n \in \mathbb{N}$ there exists $t_n \in D$ such that

$$\|f_n\| = \|f_n(t_n)\|$$

Since $(J_t)_{t\in D}$ is a countable family of weakly compact operators defined on ℓ_∞, it follows from Corollary 1.3.3 that there exists $M \in [\mathbb{N}]$ (let us recall that $[\mathbb{N}]$ is the set of all infinite subsets of $\mathbb{N}$) such that

$$J_t((\zeta_n)) = \sum_{n=1}^{\infty} \zeta_n J_t(e_n)$$

for all $(\zeta_n) \in \ell_\infty(M)$ and all $t \in D$. Of course, $\ell_\infty(M)$ is isometrically isomorphic to ℓ_∞. So, without loss of generality, we may assume that this equality holds for every $(\zeta_n) \in \ell_\infty$ and all $t \in D$. But this means that

$$J\left((\zeta_n)\right)(t) = \sum_{n=1}^{\infty} \zeta_n f_n(t) \tag{3.2}$$

whenever $(\zeta_n) \in \ell_\infty$ and $t \in D$. Let K_0 be the closure of D in K and let X_0 be the closed linear span of $\cup_{k=1}^{\infty} f_k(K_0)$. Given $(\zeta_n) \in \ell_\infty$, notice that 3.2 implies that $J\left((\zeta_n)\right)(t) \in X_0$ for all $t \in D$. Hence, by density and continuity, we have

$$J\left((\zeta_n)\right)(t) \in X_0 \tag{3.3}$$

for all $t \in K_0$. Since X_0 is a separable subspace of X, we can take now a w^*-dense sequence (x_j^*) in $B(X_0^*)$. Thanks to the Hahn-Banach theorem we can extend these functionals and assume $(x_j^*) \subset B(X^*)$. For each j let us consider the operator

$$\begin{aligned} J_j : \quad \ell_\infty &\longrightarrow \quad C(K) \\ (\zeta_n) &\longrightarrow \quad x_j^* \circ J\left((\zeta_n)\right) \end{aligned}$$

By Corollary 1.3.1, all these operators are weakly compact. And, using again Corollary 1.3.3, we deduce that there exists $M \in [\mathbb{N}]$ such that

$$J_j\left((\zeta_n)\right) = \sum_{n=1}^{\infty} \zeta_n J_j(e_n)$$

for all $(\zeta_n) \in \ell_\infty(M)$ and all $j \in \mathbb{N}$. As before, may assume $M = \mathbb{N}$, and with this in mind the preceding equality means

$$x_j^* \circ J\left((\zeta_n)\right) = \sum_{n=1}^{\infty} \zeta_n \left(x_j^* \circ J(e_n)\right)$$

for all $(\zeta_n) \in \ell_\infty$ and all $j \in \mathbb{N}$, where notice that the convergence of the series is in the norm of $C(K)$. In particular, using 3.1, it follows that given $(\zeta_n) \in \ell_\infty$ and $t \in K_0$

$$x_j^* \left(J\left((\zeta_n) \right) (t) \right) \;=\; \sum_{n=1}^{\infty} \zeta_n x_j^* \left(J(e_n)(t) \right)$$

$$=\; x_j^* \left(\sum_{n=1}^{\infty} \zeta_n J(e_n)(t) \right)$$

$$=\; x_j^* \left(\sum_{n=1}^{\infty} \zeta_n f_n(t) \right)$$

for all $j \in \mathbb{N}$. Therefore, it follows from 3.3 and the w^*-density of the x_j^*'s that

$$J\left((\zeta_n) \right)(t) = \sum_{n=1}^{\infty} \zeta_n f_n(t). \tag{3.4}$$

Now, let R be the restriction operator from $C(K, X)$ in $C(K_0, X)$ and take $T = R \circ J : \ell_\infty \to C(K_0, X)$. We have

$$\|T(e_n)\| \;=\; \|R \circ J(e_n)\| \;=\; \|R f_n\|$$
$$=\; \|f_n(t_n)\| \;=\; \|f_n\| \;=\; \|J(e_n)\| \nrightarrow 0,$$

and by 3.4, for each $(\zeta_n) \in \ell_\infty$ and each $t \in K_0$ we also have

$$T\left((\zeta_n) \right)(t) \;=\; R \circ J((\zeta_n))(t) \;=\; \sum_{n=1}^{\infty} \zeta_n R(f_n)(t)$$

$$=\; \sum_{n=1}^{\infty} \zeta_n (R \circ J)(e_n)(t) \;=\; \sum_{n=1}^{\infty} \zeta_n T(e_n)(t)$$

This is the desired contradiction and the proof is complete.

3.4 Notes and Remarks

Saab and Saab's theorem 3.1.3 is not true for $cabv(\Sigma, X)$: Talagrand in [127] provides an example of a Banach lattice which does not contain c_0 such that $cabv(\Sigma, X)$ even contains ℓ_∞.

Ferrando in [57] studies when $cabv(\Sigma, X)$ contains ℓ_∞. Recall that if the Banach space X is a dual then it contains ℓ_∞ if (and only if) it contains c_0 [35, Chapter V, Theorem 10]. Therefore, Saab & Saab's Theorem 3.1.3 may also be read as follows: "if X is a dual space then $cabv(\Sigma, X)$ contains ℓ_∞ if and only if X does". Of course, Talagrand's example just mentioned shows that the preceding result is not true for general Banach spaces.

Several results contained in this monograph have been extended to tensor products. This is specially true with Cembranos-Freniche's Theorem 3.2.1. Since it was published several extension have been given (see for instance [117]).

We have to point out that in the proof of Drewnowski's Theorem 3.3.1 some ideas introduced by Rosenthal in [110] and by Kalton in his famous paper [78] on spaces of compact operators are crucial (and part of them had already been included in section 1.6).

No good characterizations of when $C(K)$ contains a copy of ℓ_∞ or a complemented copy of c_0 are known. But we have some important partial results. For instance we have

Theorem (Rosenthal [39, Theorem VI.2.10]). *If K is extremally disconnected then $C(K)$ contains a copy of ℓ_∞.*

Recall that K is extremally disconnected (or Stonean) if the clausure of each open is open. Easy examples of extremally disconnected compact Hausdorff spaces are the Stone-Čech compactification of discrete infinite topological spaces. These spaces are very important in Banach space theory an in particular they are important in the problems we are considering. Grothendieck proved that if K is one of them then each bounded linear operator from $C(K)$ in c_0 is weakly compact [39, Corollary VI.2.12]. As an immediate consequence of this result we obtain the following

Theorem. *If K is extremally disconnected then $C(K)$ contains no complemented copies of c_0.*

The preceding result is also an easy corollary of another important fact [84, Section 11, Corollary 2]: $C(K)$ *is injective whenever K is extremally disconnected.* It is enough to observe that complemented subspaces of an injective space must be injective, while c_0 is not.

The aforementioned Grothendieck's result as well as some other related ones led to the introduction of the following definition [39, pages 156, 179]: we say that a Banach space X is Grothendieck if weak and weak* convergence of sequences in X^* coincide. Notice that thanks to the identification between w^*-null sequences and c_0-valued continuous linear operators (see Proposition 1.1.1 and [35, Chapter VII, Exercise 4]), this condition just means that each bounded linear operator from X in c_0 is weakly compact. So Grothendieck's result may be read as follows:

Theorem. *If K is extremally disconnected then $C(K)$ is a Grothendieck space.*

Clearly there are many Banach spaces with no complemented copies of c_0 which are *not* Grothendieck spaces (think for instance in ℓ_1). Nevertheless this is not the case for $C(K)$ spaces because we have

Theorem. $C(K)$ *is a Grothendieck space if and only if it does not contain any complemented copy of c_0.*

Proof. If $C(K)$ contains a complemented copy of c_0, then it can not be Grothendieck, just because clearly c_0 is not Grothendieck. Conversely, if $C(K)$ is not Grothendieck, then there exists a w^*-null non w-null sequence (μ_n) in $C(K)^*$. But, thanks to the identification between w^*-null sequences and c_0-valued continuous linear operators we have mentioned above, this means that the continuous linear operator T associated to (μ_n) is not weakly compact. So, by [39, VI.2.17.], there exists a bounded sequence (f_n) in $C(K)$ such that $f_n \cdot f_m = 0$ for $m \neq n$, and $\lim_n T(f_n) \neq 0$. It is immediate that the series $\sum f_n$ is w.u.C. (it is clear that (f_n) is even a c_0-sequence). On the other hand, it is easy to show that the condition $\lim_n T(f_n) \neq 0$ implies that there are subsequences of (μ_n) and (f_n), which we continue to denote in the same way, such that

$$\mu_n(f_n) \not\to 0.$$

So, Theorem 1.1.2 guarantees that (f_n) has a complemented c_0-subsequence. This completes the proof.

Those compacts K such that $C(K)$ is Grothendieck are some times called G-spaces. Information about these spaces may be found in [123]. For a time it was not known whether each Grothendieck $C(K)$ space must contain a copy of ℓ_∞. Haydon [68] constructed a G-space K such that $C(K)$ does not contain copies of ℓ_∞. However, this $C(K)$ does admit ℓ_∞ as a quotient. Talagrand, using continuum hypothesis, constructed a G-space K such that the corresponding $C(K)$ does not even admit ℓ_∞ as a quotient.

Extremally disconnected compacts satisfy the following two properties: (a) They do not contain copies of ℓ_∞, and (b) They do not contain complemented copies of c_0. So, one could thing there is some relationship between these two properties in $C(K)$ spaces. This is not the case. We have the following examples:

1. If $K = [0, 1]$, then $C(K) \underset{(c)}{\supset} c_0$ and $C(K) \not\supset \ell_\infty$.

2. If $K = \beta\mathbb{N} \times \beta\mathbb{N}$, then $C(K) \underset{(c)}{\supset} c_0$ and $C(K) \supset \ell_\infty$.

3. If K is any of the compact spaces given by Haydon or Talagrand, then $C(K) \underset{(c)}{\not\supset} c_0$ and $C(K) \not\supset \ell_\infty$.

4. If $K = \beta\mathbb{N}$, then $C(K) \underset{(c)}{\not\supset} c_0$ and $C(K) = \ell_\infty$.

4. $L_p(\mu, X)$ spaces

We have already studied in Chapter 2 when $L_p(\mu, X)$, for $1 \leq p < +\infty$, contains copies of c_0 or ℓ_1. We wish to consider now *complemented* copies of these spaces, and also copies of ℓ_∞ (remember that for this last space copies and complemented copies are the same). The case $p = \infty$ is postponed until next Chapter.

Let us begin with the ℓ_1 case.

4.1 When does $L_p(\mu, X)$ contain a complemented copy of ℓ_1?

The first result is a consequence of Saab & Saab's theorem 3.1.3 which provides an answer to our question in a particular case.

Theorem 4.1.1 (Bombal [6]). *Let (Ω, Σ, μ) be a finite measure space and let $1 \leq p < +\infty$. If $L_p(\mu, X)$ has a uniformly bounded complemented ℓ_1-sequence then X contains a complemented copy of ℓ_1.*

Proof. Let (f_n) be an uniformly bounded complemented ℓ_1-sequence in $L_p(\mu, X)$. By the classical perturbation theorem (see [93, Proposition 1.a.9.] or [35, Theorem 12 of Chapter V]) we may assume that the f_n's are uniformly bounded simple functions. Let us denote $f_n = \sum_i \chi_{A_i^n}(.)x_i^n$. We will assume the x_i^n's are in the unit ball of X. Let $\sum \Gamma_n$ be a w.u.C. series in $L_p(\mu, X)^*$ such that
$$\Gamma_n(f_n) = 1$$
for all $n \in \mathbb{N}$. Let Φ be the canonical map defined in remark 1.5.3, and let us denote $\Phi(\Gamma_n) = m_n$. Of course, $\sum m_n$ is a w.u.C. series in $cabv(\Sigma, X^*)$, and for each $n \in \mathbb{N}$ we have
$$\|m_n\| = \|m_n\|(\Omega) \geq \sum_i \|m_n(A_i^n)\| \geq \sum_i m_n(A_i^n)(x_i^n)$$
$$= \sum_i \Gamma_n(\chi_{A_i^n}(.)x_i^n) = \Gamma_n(f_n) = 1$$

Therefore $cabv(\Sigma, X^*)$ contains a copy of c_0. So, we deduce from E. and P. Saab's theorem 3.1.3 that X^* contains a copy of c_0, too. Of course, by Bessaga-Pełczyński's classical theorem (Theorem 10, Chapter V of [35]) this means that X contains a complemented copy of ℓ_1.

Remark 4.1.1. It is interesting to know Bombal's original proof of the preceding result [6]. It is essentially the following: Assume our measure space is $[0, 1]$ with the Lebesgue measure, and let (f_n) be a a uniformly bounded complemented ℓ_1-sequence in $L_p([0, 1], X)$. By Lusin's theorem and the classical perturbation of basis theorem, we may assume the f_n's continuous (and uniformly bounded). Considering the canonical injection from $C([0, 1], X)$ into $L_p([0, 1], X)$, it is clear that (f_n) is also a complemented ℓ_1-sequence in $C([0, 1], X)$. Now Saab & Saab's theorem 3.1.4 guarantees that X contains a complemented copy of ℓ_1, as desired.

For simplicity, we have exposed the proof for $\Omega = [0, 1]$, however we can use the same idea with any compact K instead of $[0, 1]$, and in fact, using Kakutani's theorem [40, Section 18, Theorem 2], we have the result for all finite measure spaces.

We have followed a different approach, because we have preferred to avoid such an structural theorem as Kakutani's.

Now we wish to extend Bombal's theorem to all ℓ_1-sequences. To do this we will need the following lemma.

Lemma 4.1.1. *Let (Ω, Σ, μ) be a finite measure space and let $1 < p < \infty$. If (f_n) is an ℓ_1-sequence in $L_p(\mu, X)$ then (f_n) has a normalized block sequence (f'_n) such that*

$$\sup_m \| f'_m(\omega) \| < \infty$$

for almost all $\omega \in \Omega$.

Proof. Let (f_n) be an ℓ_1-sequence in $L_p(\mu, X)$. The set $\{\| f_n(.) \| : n \in \mathbb{N}\}$ is bounded in $L_p(\mu)$ and therefore it is relatively weakly compact. So, it is relatively weakly compact in $L_1(\mu)$, too. Let $h \in L_1(\mu)$ be a w-cluster point of $\{\| f_n(.) \| : n \in \mathbb{N}\}$. Then h is in the w-closure of $\{\| f_n(.) \| : n \geq m\}$ for every $m \in \mathbb{N}$. But Mazur's theorem (see for instance Corollary 2, Ch. II of [35]) implies that for every $m \in \mathbb{N}$, h must be in the (norm) closed convex hull of $\{\| f_n(.) \| : n \geq m\}$, too. Therefore, there are an increasing sequence $1 = k_1, \dots, k_m, \dots$ of natural numbers and a sequence (ϕ_m) in $L_1(\mu)$ (norm) converging to h and such that ϕ_m is a convex combination of

$$\{\| f_{k_m}(.) \|, \dots, \| f_{k_{m+1}-1}(.) \|\}$$

for each $m \in \mathbb{N}$. Moreover, extracting a subsequence if necessary, we may suppose that (ϕ_m) converges to h almost everywhere. Hence

$$\sup_m | \phi_m(\omega) | < \infty$$

for almost all $\omega \in \Omega$. Finally, if

$$\phi_m = \sum_{i=k_m}^{k_{m+1}-1} \lambda_i \, \| f_i(.) \|,$$

with $\lambda_i \geq 0$, and $\sum_{i=k_m}^{k_{m+1}-1} \lambda_i = 1$, we define

$$g_m = \sum_{i=k_m}^{k_{m+1}-1} \lambda_i f_i \quad \text{and} \quad f'_m = \frac{g_m}{\| g_m \|_p}$$

for all $m \in \mathbb{N}$. Since (f_n) is an ℓ_1-sequence, it is clear that $\| g_m \|_p$ is bounded away from 0, and so, it follows that (f'_n) satisfies the imposed requirements.

At this moment we can answer the question of the Section.

Theorem 4.1.2 (Mendoza [97], 1992). *For $1 < p < \infty$, $L_p(\mu, X)$ contains a complemented copy of ℓ_1 if and only if X does.*

Proof. By Theorem 1.6.1, we may assume (Ω, Σ, μ) is a finite measure space.

Let us prove the non trivial implication. Let (f_n) be a complemented ℓ_1-sequence in $L_p(\mu, X)$. By the preceding lemma and well known results on block basis (see for instance 2.a.1. of [93]) we may assume

$$\sup_n \| f_n(\omega) \| < \infty \tag{4.1}$$

for almost all $\omega \in \Omega$. Since (f_n) is complemented, there is a w.u.C. series $\sum_n \Gamma_n$ in $L_p(\mu, X)^*$ such that

$$< f_n, \Gamma_n > = 1 \tag{4.2}$$

for all $n \in \mathbb{N}$. Using Theorem 1.5.4 on the representation of $L_p(\mu, X)^*$, we know that there are w^*-measurable functions $\Psi_n : \Omega \to X^*$ such that for each $n \in \mathbb{N}$, $\| \Psi_n(.) \|$ is measurable, it belongs to $\mathcal{L}_q(\mu)$ and

$$< f, \Gamma_n > = \int_\Omega < f(\omega), \Psi_n(\omega) > d\mu(\omega)$$

for all $f \in L_p(\mu, X)$. Therefore, 4.2 means that

$$\int_\Omega < f_n(\omega), \Psi_n(\omega) > d\mu(\omega) = 1 \tag{4.3}$$

for all $n \in \mathbb{N}$. Now we have two possibilities:

(a) $(< f_n(.), \Psi_n(.) >) \subset L^1(\mu)$ is not uniformly integrable. In this case, we deduce from I.2.5. of [39] that there exist $\epsilon > 0$, a sequence (A_n) of pairwise disjoint measurable subsets of Ω, and a subsequence of $(< f_n(.), \Psi_n(.) >)$, which we continue to denote in the same way, such that

$$\left| \int_{A_n} < f_n(\omega), \Psi_n(\omega) > d\mu(\omega) \right| = \left| \int_\Omega < \chi_{A_n}(\omega) f_n(\omega), \Psi_n(\omega) > d\mu(\omega) \right|$$
$$= |< \chi_{A_n}(.) f_n(.), \Gamma_n >| > \epsilon$$

for all $n \in \mathbb{N}$. Since $\sum_n \Gamma_n$ is a w.u.C. series in $L_p(\mu, X)^*$, the preceding inequality and Theorem 1.2.1 imply that $(\chi_{A_n}(.) f_n(.))$ has an ℓ_1-subsequence. But this is a contradiction because $(\chi_{A_n}(.) f_n(.))$ is a seminormalized disjointly supported sequence in $L_p(\mu, X)$ and therefore, it is an ℓ_p-sequence (Proposition 1.4.1). Thus we have to assume
(b) $(< f_n(.), \Psi_n(.) >) \subset L^1(\mu)$ is uniformly integrable. Then, for each $k \in \mathbb{N}$ put

$$\Omega_k = \{ \omega \in \Omega : \| f_n(\omega) \| \le k \ \text{ for all } n \in \mathbb{N} \}.$$

By 4.1, $\Omega \setminus \bigcup_k \Omega_k$ is a null set, hence from 4.3 and the uniform integrability of $(< f_n(.), \Psi_n(.) >)$ it follows that there exists $k_0 \in \mathbb{N}$ such that

$$\int_{\Omega_{k_0}} < f_n(\omega), \Psi_n(\omega) > d\mu(\omega) > 1/2$$

for all $n \in \mathbb{N}$. Since $\sum_n \Gamma_n$ is a w.u.C. series in $L_p(\mu, X)^*$, according to Theorem 1.2.1, $(\chi_{\Omega_{k_0}}(.) f_n(.))$ has a complemented ℓ_1-subsequence in $L_p(\mu, X)$. Now we deduce from Bombal's Theorem 4.1.1 that X has a complemented ℓ_1-sequence.

Remark 4.1.2. We would like to point out that Bourgain's theorem 2.1.1 is on the basements of all this section. In fact, Theorem 4.1.1 (and therefore, Theorem 4.1.2) lays on Saab & Saab's theorem 3.1.3, but recall that this result in turn lays on the aforementioned Bourgain's theorem.

4.2 When does $L_p(\mu, X)$ contain a copy of ℓ_∞?

The following result and its proof is strongly inspired by Drewnowski's Theorem 3.3.1 and so it borrows several Kalton's ideas. We will need again all the power of the results of Section 1.3. We will also use the preliminary work we have made in Section 1.6.

Theorem 4.2.1 (Mendoza [96], 1990). *If $1 \le p < \infty$ then $L_p(\mu, X)$ contains a copy of ℓ_∞ if and only if X does.*

Proof. The "if" part of the result is of course trivial. For the converse recall that we may assume (Ω, Σ, μ) separable and finite (Theorem 1.6.2). Let (A_k) be a sequence of measurable sets generating Σ (see the beginning of Section 1.6). Let us suppose that $L_p(\mu, X)$ contains a copy of ℓ_∞ and that X does not. Let J be an isomorphic embedding from ℓ_∞ into $L_p(\mu, X)$ and for each $k \in \mathbb{N}$ define

$$J_k : \ell_\infty \longrightarrow X$$
$$(\zeta_n) \longrightarrow \int_{A_k} J\left((\zeta_n)\right) d\mu$$

By Corollaries 1.3.1 and 1.3.3 there exists $M \in [\mathbb{N}]$ such that,

$$J_k\left((\zeta_n)\right) = \sum_{n=1}^{\infty} \zeta_n J_k(e_n)$$

for all $(\zeta_n) \in \ell_\infty(M)$ and all $k \in \mathbb{N}$. We may assume, without loss of generality, $M = \mathbb{N}$, and so we have

$$J_k\left((\zeta_n)\right) = \sum_{n=1}^{\infty} \zeta_n J_k(e_n)$$

for all $(\zeta_n) \in \ell_\infty$ and all $k \in \mathbb{N}$. Let X_0 be a closed separable subspace of X such that $J(e_n)(t) \in X_0$ μ-almost everywhere for every n. Take $(\zeta_n) \in \ell_\infty$ and $k \in \mathbb{N}$. We have

$$\int_{A_k} J\left((\zeta_n)\right) d\mu = J_k\left((\zeta_n)\right) = \sum_{n=1}^{\infty} \zeta_n J_k(e_n) = \sum_{n=1}^{\infty} \zeta_n \int_{A_k} J(e_n) d\mu.$$

Therefore,

$$\int_{A_k} J\left((\zeta_n)\right) d\mu \in X_0 \qquad \text{for all } k \in \mathbb{N} \tag{4.4}$$

On the other hand, it is straightforward to show that

$$\Sigma_0 = \{A \in \Sigma : \int_A J\left((\zeta_n)\right) d\mu \in X_0\}$$

is a sub-σ-algebra of Σ, and so, since Σ is generated by the A_k's, 4.4 means that $\Sigma_0 = \Sigma$. In other words, it follows that

$$\int_A J((\zeta_n)) d\mu \in X_0 \qquad \text{for all } A \in \Sigma.$$

Then we have (see for instance [85, X.5 Theorem 5])

$$J\left((\zeta_n)\right)(t) \in X_0 \qquad \mu\text{-almost everywhere,}$$

and so

$$J\left((\zeta_n)\right) \in L_p(\mu, X_0)$$

for every $(\zeta_n) \in \ell_\infty$. But this implies that J is an isomorphic embedding from ℓ_∞ into the separable space $L_p(\mu, X_0)$. This contradiction finishes the proof.

4.3 Complemented copies of c_0 in $L_p(\mu, X)$

In this section we wish to characterize when $L_p(\mu, X)$ contains complemented copies of c_0. For the first time, it will be important to distinguish whether the measure space (Ω, Σ, μ) is purely atomic or not.

Let us begin studying the purely atomic case.

Lemma 4.3.1. *Let* $1 \leq p < +\infty$, *let* $\sum x_k$ *be a w.u.C. series in* $\ell_p(X)$, *with* $(x_k) = ((x_{kn})_n)_k$, *and let* $(\tau_k) = ((x_{kn}^*)_n)_k$ *be a bounded sequence in* $\ell_q(X^*) = \ell_p(X)^*$ $(\frac{1}{p} + \frac{1}{q} = 1)$. *Then*

$$\lim_m \sum_{n=m}^{\infty} |x_{kn}^*(x_{kn})| = 0 \quad \text{uniformly in } k \in \mathbb{N}.$$

Proof. If the conclusion were not true, then there exist $\epsilon > 0$, an increasing sequence (m_k) of natural numbers and a subsequence of $(\tau_k(x_k))$, which we still denote in the same way, such that

$$\left| \sum_{n=m_k}^{m_{k+1}-1} x_{kn}^*(x_{kn}) \right| > \epsilon$$

for each $k \in \mathbb{N}$. Let us define the sequence $(z_k)_k = ((z_{kn})_n)_k$ in $\ell_p(X)$ by

$$z_{kn} = \begin{cases} x_{kn} & \text{if } m_k \leq n < m_{k+1} \\ 0 & \text{otherwise} \end{cases}$$

Notice that

$$|\tau_k(z_k)| = \left| \sum_{n=m_k}^{m_{k+1}-1} x_{kn}^*(x_{kn}) \right| > \epsilon.$$

Therefore the disjointly supported sequence (z_k) is seminormalized. And so (see Proposition 1.4.1), it is an ℓ_p-sequence. But Theorem 1.1.1 implies that (z_k) has a c_0-subsequence. Hence a contradiction.

Remark 4.3.1. Notice that the conclusion in the preceding lemma simply means that $\{(x_{kn}^*(x_{kn}))_n : k \in \mathbb{N}\}$ is a relatively compact subset of ℓ_1 (see [51, IV.13.3.] or [35, Exercise 6, Chapter I]).

Theorem 4.3.1 (Bombal [8], 1992). *Let* (Ω, Σ, μ) *be a purely atomic measure space and let* $1 \leq p < \infty$, *then* $L_p(\mu, X)$ *contains a complemented copy of* c_0 *if and only if* X *has the same property.*

Proof. Let us show the non trivial implication. By Lemma 1.6.1 we may assume (Ω, Σ, μ) σ-finite, and by Proposition 1.6.4, $L_p(\mu, X)$ is isometrically isomorphic to $\ell_p(X)$. So all we have to show is the following assertion: "if $\ell_p(X)$ contains a complemented copy of c_0, then so does X". Then, assume that $\ell_p(X)$ contains a complemented copy of c_0. In view of Proposition 1.1.2 there are a c_0-basis (x_k) in $\ell_p(X)$, where $x_k = (x_{kn})_n$ for each $k \in \mathbb{N}$, and a w^*-null sequence (τ_k) in $\ell_q(X^*) = \ell_p(X)^*$, with $\tau_k = (x_{kn}^*)_n$, such that

$$\tau_k(x_j) = \sum_n x_{kn}^*(x_{jn}) = \delta_{kj} \tag{4.5}$$

Notice that for each $n \in \mathbb{N}$, $\sum_k x_{kn}$ is a w.u.C. series in X and $(x_{kn}^*)_k$ is a w^*-null sequence in X^*. So, thanks to Theorem 1.1.2, the only thing we must show is that $\lim_k x_{km}^*(x_{km}) \neq 0$ for some $m \in \mathbb{N}$. But, if this is not the case, the preceding lemma implies that

$$\lim_k \tau_k(x_k) = \lim_k \sum_{n=1}^{\infty} x_{kn}^*(x_{kn}) = 0,$$

which clearly contradicts 4.5. This completes the proof.

Let us see now which is the situation in the non purely atomic case. We have the following result of Emmanuele (see also [15]):

Theorem 4.3.2 (Emmanuele [52], 1988). *Let (Ω, Σ, μ) be a non purely atomic measure space and let $1 \leq p \leq \infty$. Assume that X contains a copy of c_0, then $L_p(\mu, X)$ contains a complemented copy of c_0.*

Proof. Proposition 1.6.5 says that $L_p(\mu, X)$ contains a complemented copy of $L_p([0, 1], X)$, therefore, it is enough to show that this last space contains a complemented copy of c_0. Take a c_0-sequence (x_n) in X, let (x_n^*) be a bounded sequence in X^* such that

$$< x_n, x_m^* > = \delta_{nm}$$

for all $n, m \in \mathbb{N}$, and let (r_n) be the sequence of Rademacher functions. It is immediate that $(r_n(.)x_n)$ is a c_0-sequence in $L_p(\mu, X)$, and in fact, in example 2.1.1 we studied a slightly more complicated situation. On the other hand, let us show that $(r_n(.)x_n^*)$ is a w^*-null sequence on $L_p(\mu, X)^*$. Notice first that if f belongs to $L_1([0, 1], X)$, then

$$\lim_n \int_0^1 r_n(t) f(t) \, dt = 0 \tag{4.6}$$

This is immediate for simple functions because (r_n) is w^*-null in $L_1([0, 1])^* = L_\infty([0, 1])$, and then, by density, we can extend the result to all functions in $L_1([0, 1], X)$. Hence, for each $f \in L_p(\mu, X)$ we have

$$
\begin{aligned}
| < f, r_n(.)x_n^* > | \ &= \ \left| \int_0^1 < f(t), r_n(t)x_n^* > dt \right| \\
&= \ \left| \int_0^1 < r_n(t)f(t), x_n^* > dt \right| \\
&= \ \left| < \int_0^1 r_n(t)f(t) \ dt \ , x_n^* > \right| \\
&\leq \ \| x_n^* \| \left\| \int_0^1 r_n(t)f(t) \ dt \right\|
\end{aligned}
$$

and the last sequence goes to zero, because of 4.6. Finally, it is clear that

$$
< r_n(.)x_n, r_m(.)x_m^* > = \delta_{nm}
$$

for all $n, m \in \mathbb{N}$. So, it follows immediately from Proposition 1.1.2 that $(r_n(.)x_n)$ is complemented in $L_p(\mu, X)$, and hence the conclusion.

Of course, the preceding result leads to the following characterization.

Theorem 4.3.3. *Let (Ω, Σ, μ) be a non purely atomic measure space and let $1 \leq p < \infty$, then $L_p(\mu, X)$ contains a complemented copy of c_0 if and only if X contains a copy of c_0.*

Proof. Sufficiency is the preceding theorem and necessity is just Kwapień's Theorem 2.1.6

Finally, in the next theorem we summarize theorems 4.3.1 and 4.3.3.

Theorem 4.3.4. *Let $1 \leq p < \infty$, then $L_p(\mu, X)$ contains a complemented copy of c_0 if and only if one of the following two conditions holds:*

(a) (Ω, Σ, μ) is purely atomic and X contains a complemented copy of c_0.
(b) (Ω, Σ, μ) is not purely atomic and X contains a copy of c_0.

Remark 4.3.2. We will see in Section 5.1 that the preceding theorem is also true for $p = +\infty$ (at least for σ-finite measures μ).

5. The space $L_\infty(\mu, X)$

In the preceding chapters we have studied $C(K, X)$ spaces and $L_p(\mu, X)$ spaces for $1 \le p < +\infty$. We devote this chapter to study $L_\infty(\mu, X)$.

For this space we restrict ourselves to consider σ-finite measure spaces (Ω, Σ, μ). There are two important reasons which lead us to do this.

The first one is the very definition of $L_\infty(\mu, X)$. While for $1 \le p < \infty$ the definition of $L_p(\mu, X)$ spaces is clear for arbitrary measure spaces and most (or all) authors give the same one, for $L_\infty(\mu, X)$ the situation is completely different. There is no problem in the finite or σ-finite case, but when we leave these cases some problems arise even when X is the scalar field: for $L_\infty(\mu)$ spaces. We would like at least to attract the attention to the fact that the definitions of $L_\infty(\mu)$ given by Dunford & Schwartz [51], A. and C. Ionescu Tulcea [75] or Rudin [115] *do not* coincide.

Another important reason is that regarding the problems we are dealing with, for $L_\infty(\mu, X)$ we have a complete answer whenever the measure μ is finite (or σ-finite). However, for arbitrary measures, whichever definition of $L_\infty(\mu, X)$ one takes, only quite partial answers are known at the moment. This is because the reductions to the σ-finite case we made in Section 1.6 for $L_p(\mu, X)$ spaces, do not work for $p = \infty$ (or, at least, they do not work in an ingenuous way).

We can add that some of the definitions of $L_\infty(\mu, X)$ for arbitrary measure spaces involve some technicalities (for instance, the concept of *locally* μ-null sets) which, in our opinion, would hide the main ideas we are interested in.

Therefore, in all this chapter (Ω, Σ, μ) will be a σ-finite measure space. In the Notes and Remarks we will mention some facts which are known in more general situations.

Now, let us see which are our non trivial questions for $L_\infty(\mu, X)$.

We have already shown in Section 1.4 that $L_\infty(\mu, X)$ has many copies of ℓ_∞, and so it has many copies of c_0 and ℓ_1, moreover, it has many copies of all separable spaces just because ℓ_∞ has this property. On the other hand, since μ is σ-finite, we know that $L_\infty(\mu)$ is injective (see for instance, [37, Theorem 4.14.]), and then it can not contain complemented copies of c_0 or ℓ_1, simply because they are not injective. We could also argue that $L_\infty(\mu)$ is the dual of

$L_1(\mu)$ and it is easy to deduce from Phillips' theorem (Corollary 1.3.2) that no dual can contain a complemented copy of c_0.

Alternatively, we could say that $L_\infty(\mu)$ is isometrically isomorphic to a $C(K)$ space with K extremally disconnected (see for instance [84, Section 11, Theorem 6]), and we have considered these compact spaces in Chapter 3. We mentioned there that the corresponding $C(K)$ can not contain complemented copies of c_0. We also showed that no $C(K)$ can contain a complemented copy of ℓ_1.

Thus our goal here is to know when $L_\infty(\mu, X)$ contains complemented copies of c_0 or ℓ_1.

5.1 Complemented copies of c_0 in $L_\infty(\mu, X)$

When we studied complemented copies of c_0 in $L_p(\mu, X)$ for $1 \le p < +\infty$ (Section 4.3), we had to distinguish between purely and non purely atomic measure spaces. Now we have to do the same. Let us begin with the purely atomic case.

For a purely atomic measure, $L_\infty(\mu, X)$ is isometrically isomorphic to $\ell_\infty(X)$ (Proposition 1.6.4). Let us study this space. The first idea to do this is trying to apply the same arguments we used to study $\ell_p(X)$ for $1 \le p < +\infty$ (see Section 4.3). We will be able to apply some of them, however, it is plain that we will find extra difficulties. This is due to the fact that, likewise in the scalar case, for most purposes $\ell_\infty(X)$ is much more complicated than the others $\ell_p(X)$ spaces. This is particularly true when we deal with duals.

In view of the results of section 1.1 (see Proposition 1.1.2 and Theorem 1.1.2), to characterize when $\ell_\infty(X)$ contains a complemented copy of c_0 it is convenient to know the w^*-null sequences in $\ell_\infty(X)^*$. So, we need to work in $\ell_\infty(X)^*$ and for this reason it is worth to recall first a few important facts about the scalar case, that is, about ℓ_∞^*. We should remember that this space may be identified in a natural way with $ba(\mathbb{N})$, the space of all scalar finitely additives measures with bounded variation defined on the σ-algebra of all subsets of $\mathbb{N}$. A particularly important subspace of $ba(\mathbb{N})$ is $ca(\mathbb{N})$, the subspace of all countably additive measures. It is isometrically isomorphic to ℓ_1 and it is norm-one complemented in $\ell_\infty^* = ba(\mathbb{N})$. Thus we can decompose in a very nice way each member of $ba(\mathbb{N})$ in sum of two measures, one of them countably additive, and the other one purely finitely additive. We will see that the same can be done in $\ell_\infty(X)^*$.

It is straightforward to show that $\ell_1(X^*)$ is isometrically isomorphic in a canonical way to a subspace of $\ell_\infty(X)^*$: each $(x_n^*) \in \ell_1(X^*)$ acts as a member of $\ell_\infty(X)^*$ in the following way

$$< (x_n), (x_n^*) > = \sum_{n=1}^{\infty} < x_n, x_n^* >$$

for each $(x_n) \in \ell_\infty(X)$. So, in all this section we will consider $\ell_1(X^*)$ as a subspace of $\ell_\infty(X)^*$. Moreover, if $i_m : X \to \ell_\infty(X)$ denotes the canonical injection given by

$$i_m(x) = (\underbrace{0, \ldots, 0}_{m-1 \text{ times}}, x, 0, 0, \ldots)$$

for each $x \in X$, one can show easily that

$$P : \ell_\infty(X)^* \longrightarrow \ell_1(X^*)$$
$$\tau \longrightarrow (\tau \circ i_n)_n$$

is a well defined norm-one projection onto $\ell_1(X^*)$. This projection has an additional interesting property (see [102, Lemma 11.4] or [90, p. 53]):

Proposition 5.1.1. *The projection P defined above is w^*-sequentially continuous, that is, it transforms w^*-null sequences in w^*-null sequences (where we continue to consider $\ell_1(X^*)$ as a subspace of $\ell_\infty(X)^*$ in the canonical way).*

Proof. Let (τ_k) be a w^*-null sequence in $\ell_\infty(X)^*$. Take $x = (x_n)_n \in \ell_\infty(X)$. We must show that $< x, P(\tau_k) > \underset{k}{\to} 0$. Let us consider the continuous linear operator

$$T_x : \ell_\infty \longrightarrow \ell_\infty(X)$$
$$(t_n)_n \longrightarrow (t_n x_n)_n$$

Since (τ_k) is w^*-null in $\ell_\infty(X)^*$, it is clear that $(\tau_k \circ T_x)_k$ is w^*-null in ℓ_∞^*. Then, Phillips's lemma [35, p. 83] guarantees that

$$\lim_k \sum_{n=1}^\infty < e_n, \tau_k \circ T_x > = 0,$$

where (e_n) is the canonical unit sequence in ℓ_∞ (that is, the canonical c_0-basis). Therefore, using that

$$T_x(e_n) = (\underbrace{0, \ldots, 0}_{n-1 \text{ times}}, x, 0, 0, \ldots) = i_n(x_n)$$

we have

$$< x, P(\tau_k) > \; = \; < (x_n)_n, (\tau_k \circ i_n)_n > \; = \; \sum_{n=1}^\infty < x_n, \tau_k \circ i_n >$$

$$= \; \sum_{n=1}^\infty < e_n, \tau_k \circ T_x > \underset{k}{\longrightarrow} 0$$

Thanks to the projection P defined above, $\ell_\infty(X)^*$ is the direct sum of $\ell_1(X^*)$ (in this space we have the "countably additive part" of the functionals) and $ker(P)$, the kernel of P (in this space we have the "purely finitely additive part" of the functionals). Moreover, in view of the preceding lemma a w^*-null sequence in $\ell_\infty(X)^*$ is sum of two w^*-null sequences, one of them in $\ell_1(X^*)$ and the other one in $ker(P)$. All we have to do is to study these two kinds of w^*-null sequences. We will do this in the following two lemmas. We will see that both of them have a "good" (although different) behavior. The first lemma may be found in [89, Lemma 2.8] or [90, Lemma 4]. The second one is implicitly contained in [90].

Lemma 5.1.1. *Let* $(\tau_k) = ((x^*_{kn})_n)_k$ *be a w^*-null sequence in* $\ell_\infty(X)^*$ *which is contained in* $\ell_1(X^*)$ *, that is,*

$$\tau_k = (x^*_{kn})_n \in \ell_1(X^*)$$

for all $k \in \mathbb{N}$. Then

$$\lim_m \sum_{n=m}^\infty \|x^*_{kn}\| = 0$$

uniformly in $k \in \mathbb{N}$.

Proof. If the conclusion were not true, then there exist $\epsilon > 0$, a subsequence of (τ_k), which we still denote in the same way, and an increasing sequence (m_k) of natural numbers such that

$$\sum_{n=m_k}^{m_{k+1}-1} \|x^*_{kn}\| > \epsilon$$

for each $k \in \mathbb{N}$. Therefore, there exists a vector (x_n) in the unit ball of $\ell_\infty(X)$ such that

$$\sum_{n=m_k}^{m_{k+1}-1} <x_n, x^*_{kn}> > \epsilon \tag{5.1}$$

for each $k \in \mathbb{N}$. On the other hand, since (τ_k) is w^*-null, we have that

$$\lim_k <(t_n x_n), \tau_k> = \lim_k \sum_{n=1}^\infty <t_n x_n, x^*_{kn}> = \lim_k \sum_{n=1}^\infty t_n <x_n, x^*_{kn}> = 0$$

for each $(t_n) \in \ell_\infty$. In other words, we deduce that $((<x_n, x^*_{kn}>)_n)_k$ is a weakly null sequence in ℓ_1. Hence, we deduce from Schur theorem [35, p. 85] that $((<x_n, x^*_{kn}>)_n)_k$ is norm null, and so

$$\lim_k \sum_{n=1}^\infty |<x_n, x^*_{kn}>| = 0,$$

which contradicts 5.1. This completes the proof.

Lemma 5.1.2. *Let $\sum x_k$ be a w.u.C. series in $\ell_\infty(X)$ and assume that (τ_k) is a w^*-null sequence in $\ell_\infty(X)^*$ which lies in the kernel of P (P is the projection defined at the beginning of the Section), then*

$$\lim_k <x_k, \tau_k> = 0$$

Proof. Let $\sum x_k$ be a w.u.C. series in $\ell_\infty(X)$ and assume that (τ_k) is a w^*-null sequence in $\ell_\infty(X)^*$ which lies in the kernel of P. We will suppose that

$$<x_k, \tau_k> \not\to 0,$$

and we will reach a contradiction. Since the elements of $ker(P)$ vanish in every finitely supported vector of $\ell_\infty(X)$, we have

$$
\begin{aligned}
<x_k, \tau_k> \ &= \ < (x_{k1}, x_{k2}, \ldots, x_{kk-1}, 0, 0 \ldots), \tau_k > \\
&+ < (\underbrace{0, \ldots, 0}_{k-1 \text{ times}}, x_{kk}, x_{kk+1}, x_{kk+2}, \ldots), \tau_k > \\
&= \ < (\underbrace{0, \ldots, 0}_{k-1 \text{ times}}, x_{kk}, x_{kk+1}, x_{kk+2}, \ldots), \tau_k >
\end{aligned}
$$

Now, let us define

$$
\begin{aligned}
\Phi : \ell_\infty \ &\longrightarrow \ \ell_\infty(X) \\
(t_k) \ &\longrightarrow \ (\sum_{k=1}^{n} t_k x_{kn})_n
\end{aligned}
$$

Since $\sum x_k$ is a w.u.C. series, it is very easy to show that Φ is a well defined continuous linear operator. Notice also that if we denote by (e_k) the canonical unit sequence in ℓ_∞, then

$$\Phi(e_k) = (\underbrace{0, \ldots, 0}_{k-1 \text{ times}}, x_{kk}, x_{kk+1}, x_{kk+2}, \ldots)$$

for all $k \in \mathbb{N}$., and so

$$< e_k, \tau_k \circ \Phi > = <x_k, \tau_k> \not\to 0$$

On the other hand, $(\tau_k \circ \Phi)$ is a w^*-null sequence in ℓ_∞^*. Therefore, we deduce from Theorem 1.1.2 that (e_k) has a complemented c_0-subsequence. But this is a well known contradiction (see Corollary 1.3.2).

Theorem 5.1.1 (Leung-Räbiger, 1990). *$\ell_\infty(X)$ contains a complemented copy of c_0 if and only if X contains a complemented copy of c_0.*

Proof. Let us show the non trivial implication. Assume that $\ell_\infty(X)$ contains a complemented copy of c_0. Then, there exists a c_0-sequence (x_k) in $\ell_\infty(X)$ and a w^*-null sequence (τ_k) in $\ell_\infty(X)^*$ such that

$$< x_k, \tau_k > = 1$$

We can write

$$\tau_k = P(\tau_k) + (\tau_k - P(\tau_k)),$$

and by Proposition 5.1.1 and the preceding lemma we have

$$\lim_k < x_k, \tau_k - P(\tau_k) > = 0,$$

therefore, we conclude

$$\lim_k < x_k, P(\tau_k) > = 1 \tag{5.2}$$

Let us denote $P(\tau_k) = (x^*_{kn})_n \in \ell_1(X^*)$ and $x_k = (x_{kn})_n \in \ell_\infty(X)$. Since $\sum x_k$ is a w.u.C. series, it follows very easily that

$$\sum_k x_{nk}$$

is a w.u.C. series for every $n \in \mathbb{N}$. Besides, by the w^*-sequential continuity of P we have that $(x^*_{kn})_k$ is a w^*-null sequence in X^* for all $n \in \mathbb{N}$. On the other hand, it follows from Lemma 5.1.1 that

$$\lim_m \sum_{n=m}^{\infty} \| x^*_{kn} \| = 0$$

uniformly in $k \in \mathbb{N}$, and so

$$\lim_m \sum_{n=m}^{\infty} |< x_{kn}, x^*_{kn} >| = 0$$

uniformly in $k \in \mathbb{N}$. Therefore, we deduce from 5.2 that there exists n_0 such that

$$\lim_k < x_{kn_0}, x^*_{kn_0} > \neq 0$$

Now, it is enough a look to Theorem 1.1.2 to get the conclusion.

Of course, another way to state the preceding theorem is the following:

Theorem 5.1.2 (Leung-Räbiger, 1990). *Let (Ω, Σ, μ) be a σ-finite purely atomic measure space, then $L_\infty(\mu, X)$ contains a complemented copy of c_0 if and only if X does.*

Once we have solved the problem for purely atomic measures, let us consider the non purely atomic case. We will only need two results which are true for general $L_\infty(\mu, X)$ spaces. They were proved by Díaz [27]. The first one is a characterization of w.u.C. series in $L_\infty(\mu, X)$. Roughly speaking, it says that a series $\sum f_n$ in $L_\infty(\mu, X)$ is w.u.C. if and only if it is "(essentially) uniformly w.u.C."

Lemma 5.1.3. Let (Ω, Σ, μ) be a σ-finite measure space, then a series $\sum f_n$ in $L_\infty(\mu, X)$ is w.u.C. if and only if there exist a null subset Z of Ω and a positive number M such that

$$\left\| \sum_{n=1}^m t_n f_n(\omega) \right\| \leq M$$

for all $(t_n) \in B(\ell_\infty)$, all $m \in \mathbb{N}$ and all $\omega \in \Omega \setminus Z$.

Proof. The condition is clearly sufficient [35, Chapter V, Theorem 6]. Let us show necessity. Assume $\sum f_n$ is a w.u.C. series in $L_\infty(\mu, X)$. By [35, Chapter V, Theorem 6], there exists $M > 0$ such that

$$\left\| \sum_{n=1}^m t_n f_n \right\|_\infty \leq M$$

for all $m \in \mathbb{N}$. But the preceding inequality means that for each $m \in \mathbb{N}$ and each (t_n) belonging to $B(\ell_\infty^m)$, the unit ball of ℓ_∞^m, there exists a null subset $Z_{(m,(t_n))}$ of Ω such that

$$\left\| \sum_{n=1}^m t_n f_n(\omega) \right\| \leq M$$

for all $\omega \in \Omega \setminus Z_{(m,(t_n))}$. To get the conclusion it is enough to take $Z = \cup Z_{(m,(t_n))}$, where the union runs over the denumerable set of all $(m, (t_n))$ with $m \in \mathbb{N}$ and $(t_n) \in B(\ell_\infty^m)$ with rational coordinates.

Theorem 5.1.3 (Díaz, 1994). Let (Ω, Σ, μ) be a σ-finite measure space. If $L_\infty(\mu, X)$ contains a complemented copy of c_0, then X contains a copy of c_0.

Proof. Assume that $L_\infty(\mu, X)$ contains a complemented copy of c_0. Let (f_n) be a c_0-basis in $L_\infty(\mu, X)$ and let (τ_n) be a w^*-null sequence in $L_\infty(\mu, X)^*$ such that

$$< f_n, \tau_n > \ = \ 1$$

for all $n \in \mathbb{N}$. By the preceding lemma, there exist a null set Z of Ω and a positive number M such that

$$\left\| \sum_{n=1}^m t_n f_n(\omega) \right\| \leq M \tag{5.3}$$

for all $(t_n) \in B(\ell_\infty)$, all $m \in \mathbb{N}$ and all $\omega \in \Omega \setminus Z$. Obviously, the series $\sum f_n(\omega)$ is w.u.C. for all $\omega \in \Omega \setminus Z$. If for some of these ω's the series $\sum f_n(\omega)$ is not unconditionally convergent, then we deduce from Bessaga-Pełczyński's classical theorem [35, Chapter V, Theorem 8] that X contains a copy of c_0. Otherwise, let us show that we are led to a contradiction. We must assume that for all $\omega \in \Omega \setminus Z$ the series $\sum f_n(\omega)$ is unconditionally convergent and therefore [93, page 16],

$$\sum_{n=1}^{\infty} t_n f_n(\omega)$$

is convergent for all $(t_n) \in \ell_\infty$. Thus, if $(t_n) \in \ell_\infty$ the series $\sum t_n f_n$ is pointwise convergent almost everywhere, although it is not in general convergent in $L_\infty(\mu, X)$ because $\| f_n \|_\infty = 1$ for all n. Then, for each $(t_n) \in \ell_\infty$ we can define

$$S((t_n)) : \Omega \longrightarrow X$$
$$\omega \longrightarrow \begin{cases} \sum_{n=1}^{\infty} t_n f_n(\omega) & \text{if } \omega \in \Omega \setminus Z \\ 0 & \text{otherwise} \end{cases}$$

Now, 5.3 says that the measurable function $S((t_n))$ is bounded, and moreover, it implies that

$$S : \ell_\infty \longrightarrow L_\infty(\mu, X)$$
$$(t_n) \longrightarrow S((t_n))$$

is a well defined continuous linear operator. Therefore, $(\tau_n \circ S)$ is a w^*-null sequence in ℓ_∞^*, and, if we denote by (e_n) the canonical unit sequence in ℓ_∞, we have

$$< e_n, \tau_n \circ S > \; = \; < f_n, \tau_n > \; = \; 1$$

So, we deduce from Theorem 1.1.2 that ℓ_∞ contains a complemented copy of c_0, a contradiction (see Corollary 1.3.2) which has appeared quite often here.

Corollary 5.1.1. *If (Ω, Σ, μ) is a σ-finite non purely atomic measure space, then $L_\infty(\mu, X)$ contains a complemented copy of c_0 if and only if X contains a copy of c_0.*

Proof. Necessity is the preceding theorem. Sufficiency follows from Theorem 4.3.2

We can finally give a theorem summarizing the results of the section:

Theorem 5.1.4. *Let (Ω, Σ, μ) be a σ-finite measure space, then $L_\infty(\mu, X)$ contains a complemented copy of c_0 if and only if one of the following two conditions holds:*

(a) (Ω, Σ, μ) is purely atomic and X contains a complemented copy of c_0.

(b) (Ω, Σ, μ) is not purely atomic and X contains a copy of c_0.

Putting together the preceding theorem and Theorem 4.3.4 we get the following:

Theorem 5.1.5. *Let $1 \leq p \leq \infty$ and let (Ω, Σ, μ) be a σ-finite measure space, then we have*

1. *If the measure μ is purely atomic then $L_p(\mu, X) \underset{(c)}{\supset} c_0$ if and only if $X \underset{(c)}{\supset} c_0$.*
2. *If the measure μ is not purely atomic then $L_p(\mu, X) \underset{(c)}{\supset} c_0$ if and only if*

   $X \supset c_0$.

5.2 Complemented copies of ℓ_1 in $L_\infty(\mu, X)$

Since $L_\infty(\mu)$ is quite different to the other $L_p(\mu)$ spaces, it is not surprising we find a great difference between $L_\infty(\mu, X)$ and $L_p(\mu, X)$, too. However, taking in account the previous results in the monograph, one could expect that one of the following results be true:

(a) $L_\infty(\mu, X)$ contains a complemented copy of ℓ_1 if and only if X does.
(b) $L_\infty(\mu, X)$ contains a complemented copy of ℓ_1 if and only if X contains a copy of ℓ_1.

This is not the case. It was first noticed by Montgomery-Smith (p. 389 of [117]). He realized that an example given by Johnson [76, Remark to Theorem 1] provided a Banach space X_0 with no copies of ℓ_1 such that $L_\infty(\mu, X_0)$ contains a complemented copy of ℓ_1. Let us begin giving Johnson's example.

Example 5.2.1 (Johnson [76]). $(\sum \oplus \ell_1^n)_\infty$ contains a complemented copy of ℓ_1.

In fact, if we denote by $\mathcal{F}$ the subspace of $(\sum \oplus \ell_1^n)_\infty$ of all sequences of the form

$$((a_1), (a_1, a_2), \ldots, (a_1, a_2, \ldots, a_n), \ldots)$$

with (a_k) belonging to ℓ_1, we will show that *$\mathcal{F}$ is a 1-complemented subspace of $(\sum \oplus \ell_1^n)_\infty$ isometrically isomorphic to ℓ_1.*

It is straightforward to show that $\mathcal{F}$ is isometrically isomorphic to ℓ_1. To find a norm one projection from $(\sum \oplus \ell_1^n)_\infty$ onto $\mathcal{F}$ take a norm one continuous linear functional L on ℓ_∞ such that

$$L((a_k)) = \lim_k a_k$$

for each convergent sequence (a_k). Now, let us define

$$L_1 : (\sum \oplus \ell_1^n)_\infty \longrightarrow \mathbb{K}$$
$$((a_1^1), (a_1^2, a_2^2), \ldots, (a_1^n, \ldots, a_n^n), \ldots) \longrightarrow L((a_1^1, a_1^2, a_1^3, \ldots))$$

and for $m \geq 2$,

$$L_m : (\sum \oplus \ell_1^n)_\infty \quad \longrightarrow \quad \mathbb{K}$$
$$((a_1^1), (a_1^2, a_2^2), \ldots, (a_1^n, \ldots, a_n^n), \ldots) \quad \longrightarrow \quad L((\underbrace{0, \ldots, 0}_{m-1 \text{ times}}, a_m^m, a_m^{m+1}, a_m^{m+2}, \ldots))$$

Clearly the L_m's are norm one continuous linear functionals and for any $x = ((a_1), (a_1, a_2), \ldots, (a_1, a_2, \ldots, a_n), \ldots) \in \mathcal{F}$

$$L_m(x) = L((\underbrace{0, \ldots, 0}_{m-1 \text{ times}}, a_m, a_m, a_m, \ldots)) = a_m$$

Now for each $k \in \mathbb{N}$ define the continuous linear operator

$$P_k : (\sum \oplus \ell_1^n)_\infty \quad \longrightarrow \quad \ell_1^k$$
$$x \quad \longrightarrow \quad (L_1(x), L_2(x), \ldots, L_k(x))$$

To compute the norm of P_k take

$$x = ((a_1^1), (a_1^2, a_2^2), \ldots, (a_1^n, \ldots, a_k^n), \ldots) \in (\sum \oplus \ell_1^n)_\infty,$$

and denote $\theta_j = sgn(L_j(x))$ for $1 \leq j \leq k$. We have

$$\| P_k(x) \|_1 = \| (L_1(x), \ldots, L_k(x)) \|_1 = | L_1(x) | + \ldots + | L_k(x) |$$

$$= \theta_1 L_1(x) + \ldots + \theta_k L_k(x)$$

$$= \theta_1 L ((a_1^1, a_1^2, a_1^3, \ldots)) + \theta_2 L ((0, a_2^2, a_2^3, \ldots)) + \ldots$$
$$\ldots + \theta_k L((\underbrace{0, \ldots, 0}_{k-1 \text{ times}}, a_k^k, a_k^{k+1}, \ldots))$$

$$= L \left(\theta_1(a_1^1, a_1^2, a_1^3, \ldots) + \ldots + \theta_k(\underbrace{0, \ldots, 0}_{k-1 \text{ times}}, a_k^k, a_k^{k+1}, \ldots) \right)$$

$$\leq \| \theta_1(a_1^1, a_1^2, a_1^3, \ldots) + \theta_2(0, a_2^2, a_2^3, \ldots) + \ldots$$
$$\ldots + \theta_k(\underbrace{0, \ldots, 0}_{k-1 \text{ times}}, a_k^k, a_k^{k+1}, \ldots) \|_\infty$$

$$= \| (\theta_1 a_1^1, \theta_1 a_1^2 + \theta_2 a_2^2, \ldots, \theta_1 a_1^k + \ldots + \theta_k a_k^k, \theta_1 a_1^{k+1} + \ldots$$
$$\ldots + \theta_k a_k^{k+1}, \theta_1 a_1^{k+2} + \ldots + \theta_k a_k^{k+2}, \ldots) \|_\infty$$

$$= \sup\{| \theta_1 a_1^1 |, | \theta_1 a_1^2 + \theta_2 a_2^2 |, \ldots, | \theta_1 a_1^k + \ldots + \theta_k a_k^k |, | \theta_1 a_1^{k+1} + \ldots$$
$$\ldots + \theta_k a_k^{k+1} |, | \theta_1 a_1^{k+2} + \ldots + \theta_k a_k^{k+2} |, \ldots\}$$

$$\leq \quad \sup\{|\, a_1^1\,|, \,|\,a_1^2\,| + |\,a_2^2\,|, \ldots, \,|\,a_1^k\,| + \ldots + |\,a_k^k\,|, \,|\,a_1^{k+1}\,| + \ldots$$
$$\ldots + |\,a_k^{k+1}\,|, \,|\,a_1^{k+2}\,| + \ldots + |\,a_k^{k+2}\,|, \ldots\}$$

$$\leq \quad \sup\{\|\,(a_1^1)\,\|_1, \|\,(a_1^2, a_2^2)\,\|_1, \ldots$$
$$\ldots, \|\,(a_1^k, \ldots, a_k^k)\,\|_1, \|\,(a_1^{k+1}, \ldots, a_{k+1}^{k+1})\,\|_1, \|\,(a_1^{k+2}, \ldots, a_{k+2}^{k+2})\,\|_1, \ldots\}$$

$$= \quad \|\,x\,\|_\infty$$

Hence, we have shown that

$$\|P_k\| \leq 1$$

for all $k \in \mathbb{N}$. Now, it is immediate that

$$P : (\sum \oplus \ell_1^n)_\infty \quad \longrightarrow \quad \mathcal{F}$$
$$x \quad \longrightarrow \quad (P_1(x), P_2(x), P_3(x), \ldots)$$

is a well defined norm one projection onto $\mathcal{F}$.

Remark 5.2.1. Actually, in the preceding example it would have been natural to take as L a generalized Banach limit (see [51, II.4.22.]). In this case we could simply have taken

$$L((a_m^m, a_m^{m+1}, a_m^{m+2}, \ldots))$$

instead of

$$L((\underbrace{0, \ldots, 0}_{m-1 \text{ times}}, a_m^m, a_m^{m+1}, a_m^{m+2}, \ldots))$$

because both values coincide [51, II.4.22.].

Remark 5.2.2. In [76] Johnson mentioned Example 5.2.1 while working in very general conditions. We have preferred to study the example from a very direct and elementary point of view. So we give an explicit form of the subspace isometrically isomorphic to ℓ_1 and an explicit form of the projection, too. However we must point out that the space $(\sum \oplus \ell_1^n)_\infty$ even contains a complemented copy of $L_1([0, 1])$. This is an immediate consequence of the following result of Hagler and Stegall [67]: *A Banach space X contains a copy of $(\sum \oplus \ell_1^n)_\infty$ if and only if X^* contains a complemented copy of $L_1([0, 1])$.*

Remark 5.2.3. (Montgomery-Smith, 1992) *There exists a Banach space X_0 with no copies of ℓ_1, such that $L_\infty(\mu, X_0)$ has a complemented copy of ℓ_1.*

Consider $X_0 = (\sum \oplus \ell_1^n)_0$, that is the direct sum of $(\ell_1^n)_n$ in the sense of c_0 ([93]). On one hand it is clear that X_0 has no copies of ℓ_1 (notice for instance that $X_0^* = (\sum \oplus \ell_\infty^n)_1$ is separable). On the other hand, if we take $\ell_\infty(X_0)$, it is immediate that this space contains a complemented copy of $(\sum \oplus \ell_1^n)_\infty$, the space of the preceding example. Therefore, taking in account that $L_\infty(\mu, X_0)$ contains a complemented copy of $\ell_\infty(X_0)$ (this is a very easy fact, and it is shown in the proof of Theorem 5.2.3), we have the following chain:

$$L_\infty(\mu, X_0) \underset{(c)}{\supset} \ell_\infty(X_0) \underset{(c)}{\supset} (\sum \oplus \ell_1^n)_\infty \underset{(c)}{\supset} \ell_1$$

So, we conclude that $L_\infty(\mu, X_0)$ contains a complemented copy of ℓ_1.

Montgomery-Smith's remark was very important because it showed what one *can not* expect. However, it seemed the example involved was a very particular case and it made difficult to give a reasonable conjecture concerning which Banach spaces X provide $L_\infty(\mu, X)$ with complemented copies of ℓ_1. In fact, some people thought it was not possible to give a useful condition describing the Banach spaces X enjoying this property. For this reason, it was surprising when S. Díaz [28] realized that the crucial condition which is satisfied by the space X_0 in Montgomery-Smith's remark is quite simple. Díaz discovered that the point was that X_0 contains ℓ_1^n's uniformly complemented (see the definition below). This was a completely new point of view, because it related for the first time the theory we are considering with the local theory of Banach spaces (the theory mainly concerned with the structure of finite dimensional subspaces of the Banach spaces).

Let us give a look to a few well known and important results on the finite dimensional structure of Banach spaces. Our aim is to understand the main facts about the Banach spaces X which contain ℓ_1^n's uniformly complemented.

Given a Banach space X, we will denote by id_X the identity operator on X. If X and Y are Banach spaces and a bounded linear operator $T : X \to Y$ is an isomorphism onto $T(X)$, we will denote by T^{-1} de inverse of T, defined on $T(X)$.

For two isomorphic Banach spaces X and Y we denote by $d(X, Y)$ the Banach-Mazur distance between X and Y, that is

$$d(X, Y) = \inf\{\|T\|\,\|T^{-1}\| \colon T \text{ is an isomorphism from } X \text{ onto } Y\}.$$

Let $1 \le p \le +\infty$. We say that a Banach space X *contains ℓ_p^n's uniformly* if there exists a sequence (E_n) of finite dimensional subspaces of X such that

$$\sup_n d(E_n, \ell_p^n) < +\infty$$

If there are also projections P_n from X onto E_n with

$$\sup_n \|P_n\| < +\infty$$

we say that X *contains ℓ_p^n's uniformly complemented.* Thanks to the injectivity of all ℓ_∞^n's it is clear that in the case $p = +\infty$ both definitions coincide.

Notice that another way of saying that X contains ℓ_p^n's uniformly complemented is to say that for each $n \in \mathbb{N}$ there are bounded linear operators

$$\ell_p^n \xrightarrow{I_n} X \xrightarrow{Q_n} \ell_p^n$$

satisfying

$$Q_n \circ I_n = id_{\ell_p^n} \quad \text{and} \quad \sup_n \|Q_n\| \, \|I_n\| < +\infty$$

If one takes adjoints, it is clear that if X contains ℓ_p^n's uniformly complemented then X^* contains ℓ_q^n's uniformly complemented, where, as usual, q is the conjugate of p, that is, $\frac{1}{p} + \frac{1}{q} = 1$. Therefore, applying the argument again, we deduce that X^{**} contains ℓ_p^n's uniformly complemented, too. The converse is also true, that is, if X^{**} contains ℓ_p^n's uniformly complemented then X has the same property. This follows easily from the following version of the Local Reflexivity Principle [92, Theorem II.5.1]:

Theorem 5.2.1 (Local Reflexivity Principle). *Let X be a Banach space and let E, G be finite dimensional subspaces of X^{**} and X^*, respectively, and let $\epsilon > 0$. Assume that P is a projection from X^{**} onto E. Then there are an injective linear operator $T : E \longrightarrow X$ and a projection P_0 from X onto $T(E)$ such that*

1. $T \mid_{E \cap X} = id_{E \cap X}$.
2. $< T(x^{**}), x^* > \, = \, < x^*, x^{**} >$ *for all $x^{**} \in E$ and all $x^* \in G$.*
3. $\|T\| \, \|T^{-1}\| \le 1 + \epsilon$ *(and so we can assume $\|T\| \le 1 + \epsilon$), and*
4. $\|P_0\| \le (1 + \epsilon) \, \|P\|$.

The following proposition summarizes our remarks given above.

Proposition 5.2.1. *Let X be a Banach space, let $1 \le p \le +\infty$ and let us denote by q the conjugate of p (that is, $\frac{1}{p} + \frac{1}{q} = 1$). Then the following are equivalent:*

(a) *X contains ℓ_p^n's uniformly complemented.*
(b) *X^* contains ℓ_q^n's uniformly complemented.*
(c) *X^{**} contains ℓ_p^n's uniformly complemented.*

Now, let us concentrate our attention in the case $p = 1$ (and by duality, in the case $p = +\infty$, too). We need a good characterization of the Banach spaces containing ℓ_1^n's uniformly complemented. This will follow from the following particular version of the "Great Theorem" of Maurey-Pisier (remember that the "only if" part in the theorem is trivial, the difficult part is the converse):

Theorem 5.2.2 (Maurey-Pisier, Theorem 14.1 of [37]). *X contains ℓ_∞^n's uniformly if and only if X does not have finite cotype.*

With the two preceding results in mind the following characterization follows immediately. As usual, let us denote by $c_0(X)$ the Banach space of all null sequences in X, endowed with the supremum norm. We only have to recall that $\ell_1(X^*)$ is the dual of $c_0(X)$ and that X and $\ell_1(X)$ has the same cotype [37, Theorem 11.12].

Proposition 5.2.2. *The following are equivalent:*

(a) X contains ℓ_1^n's uniformly complemented.
(b) $c_0(X)$ contains ℓ_1^n's uniformly complemented.
(c) X^* does not have finite cotype.
(d) $\ell_1(X^*)$ does not have finite cotype.

Once we know the main facts concerning Banach spaces which contain ℓ_1^n's uniformly complemented, we need two more ingredients to get the characterization of Banach spaces X such that $L_\infty(\mu, X)$ contains a complemented copy of ℓ_1. The first one is a very well known perturbation lemma. We will include the proof for the sake of completeness. As a consequence of it we will show the close relationship between the finite-dimensional subspaces of $L_\infty(\mu, X)$ and $\ell_\infty(X)$. The second ingredient is Kalton's contribution. It is a lemma which relates $\ell_\infty(X^{**})$ and $\ell_\infty(X)^{**}$. It is crucial in the proof of the main theorem 5.2.3 (see also the Notes and Remarks).

Lemma 5.2.1 (Perturbation lemma). *Let F be a finite dimensional subspace of X and let P be a projection from X onto F. Suppose $dim(F) = n$ and let $\{x_1, \ldots, x_n\}$ be a unitary basis of F. Then, given $\epsilon > 0$ there exists $\delta > 0$ such that if $\{y_1, \ldots, y_n\}$ are vectors in X satisfying*

$$\max_{1 \le i \le n} \| x_i - y_i \| < \delta,$$

and we denote by G the linear span of the y_i's, we have

(i) $d(F, G) \le 1 + \epsilon$, and
(ii) There exists a projection P_o from X onto G with $\| P_o \| \le (1 + \epsilon) \| P \|$.

Proof. First, remember that if Y is a subspace of X and $T : X \to Y$ is a bounded linear operator which satisfies $\| id_Y - T \| \leq \eta < 1$, then T is an isomorphism onto its range $T(Y)$. Moreover, since for any $y \in Y$

$$\| T(y) \| \leq \| y - T(y) \| + \| y \| \leq (1 + \eta) \| y \|, \quad \text{and}$$
$$\| T(y) \| \geq \| y \| - \| y - T(y) \| \geq (1 - \eta) \| y \|,$$

it follows that

$$\| T \| \, \| T^{-1} \| \leq \frac{1 + \eta}{1 - \eta}$$

Suppose now the assumptions of our lemma are satisfied. Then, there exists $M > 0$ such that

$$M \sum_{i=1}^{n} | \lambda_i | \leq \| \sum_{i=1}^{n} \lambda_i x_i \|$$

for all scalars (λ_i). Given $\epsilon > 0$ take $\eta \in (0, 1)$ such that

$$\frac{1 + \eta}{1 - \eta} \leq 1 + \epsilon$$

Let us see that

$$\delta = \frac{\eta}{M(\| P \| + 1)}$$

satisfies the required properties.

Take $\{y_1, \ldots, y_n\}$ vectors in X verifying

$$\max_{1 \leq i \leq n} \| x_i - y_i \| < \delta,$$

and denote by G the linear span of the y_i's. For $x = \sum_{i=1}^{n} \lambda_i x_i \in F$, define $S(x) = \sum_{i=1}^{n} \lambda_i y_i \in G$. Since

$$\| x - S(x) \| = \| \sum_{i=1}^{n} \lambda_i x_i - \sum_{i=1}^{n} \lambda_i y_i \| \leq \sum_{i=1}^{n} | \lambda_i | \, \| x_i - y_i \| \leq \delta \sum_{i=1}^{n} | \lambda_i |$$
$$\leq \delta M \| x \|,$$

we have

$$\| id_F - S \| \leq \delta M = \frac{\eta}{(\| P \| + 1)} < \eta < 1$$

Therefore, S is an isomorphism from F onto G with

$$\| S \| \, \| S^{-1} \| \leq \frac{1 + \eta}{1 - \eta} \leq 1 + \epsilon$$

Thus, $d(F, G) \leq \| S \| \, \| S^{-1} \| \leq 1 + \epsilon$.

In order to get the projection P_0, consider the operator $T : X \to X$ defined by $T = id_X + (S \circ P) - P$. It verifies

$$\| T - id_X \| = \| (S - id_F) \circ P \| \le \frac{\eta}{(\| P \| + 1)} \| P \| < \eta < 1.$$

Hence, it is well known that T is an automorphism on X. Moreover,

$$\| T \| \| T^{-1} \| \le \frac{1 + \eta}{1 - \eta} < 1 + \epsilon.$$

Since $T(F) = G$, it is clear that $P_0 = T \circ P \circ T^{-1}$ is a projection from X onto G with $\| P_0 \| \le (1 + \epsilon) \| P \|$, as required.

Proposition 5.2.3. Let (Ω, Σ, μ) be a σ-finite measure space, let $\mathcal{F}$ be a finite dimensional subspace of $L_\infty(\mu, X)$ and let P be a projection from $L_\infty(\mu, X)$ onto $\mathcal{F}$. Then, given $\epsilon > 0$ there exist a finite dimensional subspace $\mathcal{G}$ of $\ell_\infty(X)$ and a projection P_0 from $\ell_\infty(X)$ onto $\mathcal{G}$ such that

$$d(\mathcal{F}, \mathcal{G}) \le 1 + \epsilon \quad and \quad \| P_0 \| \le (1 + \epsilon) \| P \|$$

Proof. Assume the hypothesis of the proposition holds, let $\epsilon > 0$, and apply the preceding lemma to the finite dimensional subspace $\mathcal{F}$ of $L_\infty(\mu, X)$, taking any unitary basis $\{f_1, \dots, f_n\}$ of $\mathcal{F}$. Given the positive number δ provided by the lemma, take $\{g_1, \dots, g_n\}$ in $L_\infty(\mu, X)$ *countably valued* such that

$$\max_{1 \le i \le n} \| f_i - g_i \|_\infty < \delta$$

Let (A_m) be a sequence of pairwise disjoint measurable sets with positive measure such that each g_i is constant in the A_m's and $\mu(\Omega \setminus \cup_m A_m) = 0$. Let us denote by $\mathcal{H}$ the subspace of $L_\infty(\mu, X)$ of all functions which are constant in the A_m's. Of course, $\mathcal{H}$ is isometrically isomorphic to $\ell_\infty(X)$. If we denote by $\mathcal{G}$ the finite dimensional subspace of $L_\infty(\mu, X)$ spanned by the g_i's, we have

$$\mathcal{G} \subset \mathcal{H} \subset L_\infty(\mu, X),$$

and the preceding lemma guarantees that

$$d(\mathcal{F}, \mathcal{G}) \le 1 + \epsilon,$$

and that there is a projection P_0 from $L_\infty(\mu, X)$ onto $\mathcal{G}$ such that

$$\| P_0 \| \le (1 + \epsilon) \| P \| .$$

If we consider the restriction of P_0 to $\mathcal{H}$, we get the desired conclusion.

As usual, in the next lemma we identify in the canonical way X as a subspace of X^{**}. In this way $\ell_\infty(X)$ is a subspace of $\ell_\infty(X^{**})$ and a subspace of $\ell_\infty(X)^{**}$, too.

Lemma 5.2.2 (Kalton). *Given $\epsilon > 0$ there exists a continuous linear operator*

$$J : \ell_\infty(X^{**}) \longrightarrow \ell_\infty(X)^{**}$$

such that $\| J \| \leq 1 + \epsilon$ and $J|_{\ell_\infty(X)} = id_{\ell_\infty(X)}$.

Proof. Let $\mathfrak{F}$ be a finite dimensional subspace of $\ell_\infty(X^{**})$. Then there exists a sequence (F_n) of finite dimensional subspaces of X^{**} such that

$$\mathfrak{F} \subset \left(\sum \oplus F_n\right)_\infty \subset \ell_\infty(X^{**}).$$

By the local reflexivity principle (Theorem 5.2.1), for each $n \in \mathbb{N}$ there exists an injective operator $u_n : F_n \to X$ such that

$$u_n|_{X \cap F_n} = id_{X \cap F_n} \quad \text{and} \quad \| u_n \| < 1 + \epsilon.$$

Now we can define

$$
\begin{aligned}
I_{\mathfrak{F}} : \mathfrak{F} &\longrightarrow \ell_\infty(X) \\
(x_n) &\longrightarrow (u_n(x_n))
\end{aligned}
$$

It is clear that $I_{\mathfrak{F}}$ is a well defined continuous linear operator which satisfies

$$\| I_{\mathfrak{F}} \| \leq 1 + \epsilon \quad \text{and} \quad I_{\mathfrak{F}}|_{\mathfrak{F} \cap \ell_\infty(X)} = id_{\mathfrak{F} \cap \ell_\infty(X)},$$

Now we will use a classical Lindenstrauss compactness argument. Denote by $B(\ell_\infty(X^{**}))$ and $B(\ell_\infty(X)^{**})$ the unit balls of $\ell_\infty(X^{**})$ and $\ell_\infty(X)^{**}$ respectively, and define

$$
\begin{aligned}
J_{\mathfrak{F}} : B(\ell_\infty(X^{**})) &\longrightarrow \ell_\infty(X) \\
x &\longrightarrow \begin{cases} I_{\mathfrak{F}}(x) & \text{if } x \in \mathfrak{F} \\ 0 & \text{otherwise} \end{cases}
\end{aligned}
$$

Of course $J_{\mathfrak{F}}$ neither is linear nor continuous, but notice that

$$J_{\mathfrak{F}}(B(\ell_\infty(X^{**}))) \subset (1 + \epsilon)B(\ell_\infty(X)) \subset (1 + \epsilon)B(\ell_\infty(X)^{**}) \tag{5.4}$$

Now we can consider the net $(J_{\mathfrak{F}})_{\mathfrak{F}}$, where $\mathfrak{F}$ runs on the directed set of all finite dimensional subspaces of $\ell_\infty(X^{**})$. By 5.4, this is a net in the compact space

$$\left((1 + \epsilon)B(\ell_\infty(X)^{**}), w^* \right)^{B(\ell_\infty(X^{**}))}.$$

Therefore $(J_{\mathfrak{F}})_{\mathfrak{F}}$ has an accumulation point J in this compact space. It is immediate that J can be extended to a linear operator, which we continue to denote J, from the whole space $\ell_\infty(X^{**})$ into $\ell_\infty(X)^{**}$, and it is also immediate that this linear operator J has the required properties.

We can finally state and prove the main result of the Section.

Theorem 5.2.3 (Díaz-Kalton). *Let (Ω, Σ, μ) be a σ-finite measure space, then the following are equivalent:*

(a) $L_\infty(\mu, X)$ contains a complemented copy of ℓ_1.
(b) $L_\infty(\mu, X)$ contains ℓ_1^n's uniformly complemented.
(c) $\ell_\infty(X)$ contains a complemented copy of ℓ_1.
(d) $\ell_\infty(X)$ contains ℓ_1^n's uniformly complemented.
(e) X contains ℓ_1^n's uniformly complemented.

Proof. We will show the following chain of implications: $(a) \Rightarrow (b) \Rightarrow (d) \Rightarrow (e) \Rightarrow (c) \Rightarrow (a)$.

That (a) implies (b) is trivial, and (b) implies (d) is an immediate consequence of Proposition 5.2.3.

Let us show now that (d) implies (e). We assume that $\ell_\infty(X)$ contains ℓ_1^n's uniformly complemented. So there exist $M > 0$, a sequence (F_n) of finite dimensional subspaces of $\ell_\infty(X)$ and a sequence of projections

$$P_n : \ell_\infty(X) \to F_n$$

such that

$$d(F_n, \ell_1^n) \le M \quad \text{and} \quad \| P_n \| \le M$$

for all $n \in \mathbb{N}$. Kalton's lemma 5.2.2 provides a continuous linear operator

$$J : \ell_\infty(X^{**}) \longrightarrow \ell_\infty(X)^{**}$$

such that

$$J|_{\ell_\infty(X)} = id_{\ell_\infty(X)}$$

Of course we may consider the F_n's as subspaces of $\ell_\infty(X^{**})$, and if we denote by P_n^{**} the second adjoint of P_n, it is clear that the operators

$$P_n^{**} \circ J : \ell_\infty(X^{**}) \to F_n$$

provide a bounded sequence of projections. Therefore, $\ell_\infty(X^{**}) = c_0(X)^{**}$ contains ℓ_1^n's uniformly complemented. Then, by Proposition 5.2.1 this means that $c_0(X)$ contains ℓ_1^n's uniformly complemented, too. But now, thanks to Proposition 5.2.2, we deduce that (e) holds.

To see that (e) implies (c) is very easy. Assume that there exist $M > 0$, a sequence (F_n) of finite dimensional subspaces of X and a sequence (P_n) of projections from X onto F_n such that

$$d(F_n, \ell_1^n) \le M \quad \text{and} \quad \| P_n \| \le M$$

for all $n \in \mathbb{N}$. It is clear that $(\sum \oplus F_n)_\infty$ is isomorphic to $(\sum \oplus \ell_1^n)_\infty$, and we know (Johnson's example 5.2.1) that $(\sum \oplus \ell_1^n)_\infty$ contains a complemented copy of ℓ_1. Therefore $(\sum \oplus F_n)_\infty$ contains a complemented copy of ℓ_1, too. On the other hand, it is clear that

$$(P_n) : \ell_\infty(X) \longrightarrow (\sum \oplus F_n)_\infty$$
$$x \longrightarrow (P_n(x))$$

is a projection onto $(\sum \oplus F_n)_\infty$. So, it follows that $\ell_\infty(X)$ contains a complemented copy of ℓ_1.

Finally, note that $\ell_\infty(X)$ is isometrically isomorphic to a complemented subspace of $L_\infty(\mu, X)$. Indeed, likewise in the proof of proposition 5.2.3, take a sequence (A_m) of pairwise disjoint measurable sets with finite and positive measure, and consider the subspace $\mathcal{H}$ of $L_\infty(\mu, X)$ of all functions which are constant in the A_m's. Then the map

$$L_\infty(\mu, X) \longrightarrow \mathcal{H}$$
$$f \longrightarrow \sum_n \chi_{A_n}(\cdot) \frac{1}{\mu(A_n)} \Big(\int_{A_n} f d\mu \Big)$$

is a projection (it is a conditional expectation). Since $\mathcal{H}$ is clearly isometrically isomorphic to $\ell_\infty(X)$, our statement is proved. Hence (c) implies (a), obviously. This completes the proof.

Remark 5.2.4. It should be pointed out that the preceding result provides immediately examples of Banach spaces X which are even separable and reflexive for which $L_\infty(\mu, X)$ contains a complemented copy of ℓ_1. Take for instance $X = (\sum \oplus \ell_1^n)_2$.

5.3 Notes and Remarks

As we have already mentioned, in this chapter we have considered only σ-finite measure spaces, while in the rest of the monograph we have been working in arbitrary measure spaces. In the introduction of the chapter we have explained why, but we would like to call the attention here to a significant fact. At the moment, Banach space specialists usually study almost exclusively $L_\infty(\mu)$ spaces for finite (or σ-finite) measures μ, with only one important exception: spaces of type $\ell_\infty(\Gamma)$. Here, as usual, $\ell_\infty(\Gamma)$ denotes the Banach space of all bounded scalar functions defined on a certain set Γ, endowed with the supremum norm. The vectorial version of this space is $\ell_\infty(\Gamma, X)$, which is defined in the natural way.

We believe that while considering the finite measure case, we have exposed most of the main ideas related with our problems, but we would like to mention also the results which are known in more general situations.

Concerning the search of complemented copies of c_0 in $L_\infty(\mu, X)$ there are two results. One for the purely atomic case, and the other one for the non purely atomic one:

(a) Leung and Räbiger [90] showed that if the set Γ of indices is measurable (which includes of course the denumerable case), then $\ell_\infty(\Gamma, X)$ contains a complemented copy of c_0 if and only if X does. They left open the general problem for arbitrary sets of indices.

(b) Díaz proved in [27] his theorem 5.1.3 (and corollary 5.1.1) for arbitrary measure spaces.

About complemented copies of ℓ_1 in $L_\infty(\mu, X)$ we can say that if X contains ℓ_1^n's uniformly complemented then $L_\infty(\mu, X)$ contains a complemented copy of ℓ_1 (this is just because if X contains ℓ_1^n's uniformly complemented, by theorem 5.2.3, $\ell_\infty(X)$ contains a complemented copy of ℓ_1, and this space is clearly complemented in $L_\infty(\mu, X)$). However, we do not know if the converse is true or not.

We must say that actually Leung and Räbiger [90] work in spaces more general than $\ell_\infty(\Gamma, X)$. They study sums in the sense of ℓ_∞ of families of Banach spaces. However, we would add that, concerning the problem we are dealing with, these two kind of spaces present a completely analogous behavior.

Now, let us give a look at the results of the chapter.

At the beginning of Section 5.1 we recall that we can decompose in a very nice way each member of $ba(\mathbb{N}) = \ell_\infty^*$ in sum of two measures, one of them countably additive, and the other one purely finitely additive. Of course, this is just a particular case of Yosida-Hewitt decomposition theorem [51, III.5.8].

When we move the above decomposition from ℓ_∞^* to $\ell_\infty(X)^*$ we introduce the projection P in $\ell_\infty(X)^*$ which appears at the beginning of the section, and which is crucial to get Leung and Räbiger' theorem 5.1.1. It is worth to observe that this projection could also be viewed from a more general point of view. One can easily show that if X is a closed subspace of Y and there exists a bounded linear operator $S : Y \to X^{**}$ such that $S|_X = id_X$ (where, as usual, we identify X with a closed subspace of X^{**} in the canonical way), then there exists a continuous linear projection $P : Y^* \to Y^*$ whose kernel is $X^\perp = \{x^* \in Y^* : x^*(X) = \{0\}\}$, and therefore, $X^* = Y^*/X^\perp$ is complemented in Y^*. In particular, our assumption is satisfied whenever $X \subset Y \subset X^{**}$, taking as S the natural embedding. If we apply this to

$$c_0(X) \subset \ell_\infty(X) \subset \ell_\infty(X^{**}) = c_0(X)^{**},$$

we get a projection P from $\ell_\infty(X)^*$ onto its subspace $c_0(X)^* = \ell_1(X^*)$. However, we do *not* conclude from the general theory that this projection is w^*-sequentially continuous, which is very important for us. This is true in our particular situation, as we show in Proposition 5.1.1.

In this monograph we have hardly dealt with $(\sum \oplus X_n)_p$ spaces, but one should notice that Johnson's example also shows that the behavior of $(\sum \oplus X_n)_\infty$ is quite different to the one of $(\sum \oplus X_n)_p$ for $1 \le p < +\infty$.

We would like to point out that the classical perturbation lemma about basis ([93, 1.a.9.] or [35, Chapter V, Theorem 12]) is just an infinite-dimensional version of the well known lemma 5.2.1.

In Section 5.2, we explain in general when we say that a Banach space X contains ℓ_p^n's uniformly (complemented) for $1 \le p \le +\infty$. Actually we are only interested in the case $p = 1$ (and by duality, in the case $p = +\infty$), but we believe that to consider the general situation helps to understand better these concepts.

Notice that Theorem 5.2.3, the main result of Section 2, lies on the "Great Theorem" of Maurey-Pisier (Theorem 5.2.2). This is because this theorem provides a good characterization of when X contains ℓ_∞^n's uniformly complemented, and by duality and the Local Reflexivity Principle (Theorem 5.2.1) this provides a good characterization of when X contains ℓ_1^n's uniformly complemented. We do not know of any "elementary proof".

Theorem 5.2.3 is due to S. Díaz [28] and N.J. Kalton. S. Díaz in [28] proved that conditions (a), (b), (c) and (d) are equivalent, and that they hold whenever (e) is satisfied. He also showed that the five condition (a), (b), (c), (d) and (e) are equivalent in quite general conditions (for instance, for Banach lattices). Therefore, the only open problem left by him was whether (a) (or any of the other equivalent conditions (b), (c) or (d)) implied (e) or not.

In the Spring of 1996 we asked N.J. Kalton about this. He showed us how to prove that (d) indeed implies (e), and kindly allowed us to include his solution here. We have summarized it in lemma 5.2.2 which is a reformulation of Kalton's ideas contained in [79]. We believe that the lemma is much more subtle than it seems and provides a very interesting connection between $\ell_\infty(X^{**})$ and $\ell_\infty(X)^{**}$. In fact, we have seen several times people proclaiming that necessity of condition (e) was very easy and later on they were not able to give a proof. We may add that we have read several wrong proofs of the assertion (some of these wrong proofs were made by us). To show necessity of (e) in Kalton-Díaz Theorem 5.2.3 one follows more or less the following idea: if $\ell_\infty(X)$ contains ℓ_1^n's uniformly complemented, then its second dual, $\ell_\infty(X)^{**}$, does either. From this fact we wish to conclude that $\ell_\infty(X^{**}) = \ell_1(X^*)^* = c_0(X)^{**}$ has the same property, but these spaces are quite well known, and thanks to Maurey-Pisier theorem we may conclude our thesis. The point is that although $\ell_\infty(X)^{**}$ and $\ell_\infty(X^{**})$ might seem close in a first view, a careful look shows us that they are indeed very different. This is the problem solved by Kalton's lemma.

We wish to point out that Kalton's lemma is closely related with the important notion of *locally complemented subspace*. From this point of view it just says that $\ell_\infty(X)$ is *locally complemented* in $\ell_\infty(X^{**})$. The notion of locally complemented subspace has been studied by several authors and has found many interesting applications in Banach space theory (see [55], [79] and [64]). Its isometric version has been considered recently in [91] and [106].

It is worth noticing that Díaz-Kalton Theorem 5.2.3 and Hagler-Stegall result mentioned in remark 5.2.2 can be combined fruitfully to obtain a significant consequence:

Theorem. *Let (Ω, Σ, μ) be a σ-finite measure space, then the following are equivalent:*

(a) $L_\infty(\mu, X)$ *contains a complemented copy of* $L_1([0,1])$.
(b) $\ell_\infty(X)$ *contains a complemented copy of* $L_1([0,1])$.
(c) $L_\infty(\mu, X)$ *contains a complemented copy of* ℓ_1.
(d) $\ell_\infty(X)$ *contains a complemented copy of* ℓ_1.
(e) X *contains* ℓ_1^n's *uniformly complemented.*

Proof. We know by Theorem 5.2.3 that (c), (d) and (e) are equivalent. On the other hand, $\ell_\infty(X)$ contains a complemented copy of $(\sum \oplus \ell_1^n)_\infty$ whenever (e) holds, as shown in the proof of Díaz-Kalton Theorem 5.2.3. Thus, Hagler-Stegall's result mentioned in remark 5.2.2 proves that (e) implies (b). To see that (b) implies (a), it is enough to observe that $L_\infty(\mu, X)$ contains a complemented copy of $\ell_\infty(X)$ (this easy fact was shown in the proof of Theorem 5.2.3). Finally, (a) implies (c) is trivial.

Díaz [28] has also proved that *for* $1 \leq p < +\infty$, $L_\infty(\mu, X)$ *contains a complemented copy of* ℓ_p *whenever* X *contains* ℓ_p^n's *uniformly complemented.*

6. Tabulation of Results

Here is a table which summarizes all the results of our monograph.

	$C(K, X)$	$L_1(\mu, X)$	$L_p(\mu, X)$ $1 < p < +\infty$	$L_\infty(\mu, X)$
$\supset c_0$	always (trivial)	$X \supset c_0$ (2.1.6)	$X \supset c_0$ (2.1.6)	always (trivial)
$\supset c_0$ (c)	always (3.2.1)	$X \supset c_0$ (c) or $X \supset c_0$ and μ is not purely atomic (4.3.4)	$X \supset c_0$ (c) or $X \supset c_0$ and μ is not purely atomic (4.3.4)	$X \supset c_0$ (c) or $X \supset c_0$ and μ is not purely atomic (5.1.4)
$\supset \ell_1$	$C(K) \supset \ell_1$ or $X \supset \ell_1$ (3.1.2)	always (trivial)	$X \supset \ell_1$ (2.2.2)	always (trivial)
$\supset \ell_1$ (c)	$X \supset \ell_1$ (c) (3.1.4)	always (trivial)	$X \supset \ell_1$ (c) (4.1.2)	$X \supset \ell_1^n$ uniformly complemented (5.2.3)
$\supset \ell_\infty$	$C(K) \supset \ell_\infty$ or $X \supset \ell_\infty$ (3.3.1)	$X \supset \ell_\infty$ (4.2.1)	$X \supset \ell_\infty$ (4.2.1)	always (trivial)

The meaning of the Table is clear. In the first row we have the spaces we are studying, in the first column we see the different containments we are considering, and in each box we put the condition which must be satisfied and the number of the corresponding theorem if it is not trivial. "Trivial" means that it follows from the following easy containments

$$C(K, X) \underset{(c)}{\supset} C(K) \supset c_0 \quad L_1(\mu, X) \underset{(c)}{\supset} L_1(\mu) \underset{(c)}{\supset} \ell_1$$

$$L_\infty(\mu, X) \underset{(c)}{\supset} L_\infty(\mu) \underset{(c)}{\supset} \ell_\infty \supset c_0, \ell_1$$

Recall that we are assuming $C(K)$, $L_p(\mu)$ and X infinite-dimensional.

For $1 \leq p < +\infty$ the measure μ in $L_p(\mu, X)$ spaces is arbitrary, in the case $p = +\infty$ it is assumed to be σ-finite.

A look at the table shows that in most cases we get the natural answers, that is, we get affirmative answers to Problem 2 in the Introduction. There are only the following exceptions: (a) The different containments of c_0 as a complemented subspace, and (b) Complemented copies of ℓ_1 in $L_\infty(\mu, X)$. In both cases we see that vector valued function spaces contain complemented copies of c_0 or ℓ_1 more often than expected. Notice once more that case (b) provides the only answer which depends exclusively on the local structure of the Banach space X.

7. Some Related Open Problems

To finish our work we would like to mention a few lines of research and open problems which are closely related with the results of this monograph.

I. If we remember Problem 1 posed in the Introduction, we see that we have a complete solution to it with only one exception: For $L_\infty(\mu, X)$ we know the answer for σ-finite spaces, but for arbitrary positive measure spaces we do not. Then we have the following

Problem 7.1. *Assume (Ω, Σ, μ) is a non σ-finite measure space, when does $L_\infty(\mu, X)$ have a complemented copy of c_0?, when does it have a complemented copy of ℓ_1?*

It would be particularly interesting to answer these questions for the space $L_\infty(\mu, X) = \ell_\infty(\Gamma, X)$.

Leung and Räbiger [90] already posed this problem for complemented copies of c_0 in the sum in the sense of ℓ_∞ of an arbitrary family $\{X_i : i \in \Gamma\}$ of Banach spaces. See Section 5.3, and in general Chapter 5.

II. In order to consider analogous problems to the ones studied here, a natural thing is to add to our list of spaces (c_0, ℓ_1 and ℓ_∞) the ℓ_r spaces ($1 < r < \infty$), that is, we can ask:

Problem 7.2. *When does $C(K, X)$ or $L_p(\mu, X)$ have copies or complemented copies of ℓ_r ($1 < r < \infty$)?*

The general problem seems to be very difficult because no good characterization of spaces containing ℓ_r ($1 < r < \infty$) is known. However, some partial answers have already been given (see [11], [28], [71], [107], [108], [119], [120]). Let us mention at least that $L_p(\mu, X)$ can contain copies of ℓ_r, while neither $L_p(\mu)$ nor X contain a copy of ℓ_r (this is an old known result, see [107], [108] or [94, Remark to Lemma 1.f.8]).

And in the same way we can consider $L_1([0, 1])$:

Problem 7.3. *If $L_p(\mu, X)$ $(1 < p < \infty)$ contains a copy of $L_1([0, 1])$ must X contain a copy of $L_1([0, 1])$? or, in general, when does $L_p(\mu, X)$ have a copy of $L_1([0, 1])$?*

These questions were posed to the authors by J. Diestel and they can also be found in [117, Question 10]. The answer to the first one is yes if X is a dual space [117], but we do not know a complete answer even for $p = 2$.

What about *complemented* copies of $L_1([0, 1])$? In this case the question is natural for $L_\infty(\mu, X)$ and $C(K, X)$, too. We have just shown in the Notes and Remarks of Chapter 5 a complete characterization for $L_\infty(\mu, X)$. But concerning the other spaces we do not know anything.

Problem 7.4. *When does $C(K, X)$ or $L_p(\mu, X)$ $(1 < p < \infty)$ contain a complemented copy of $L_1([0, 1])$?*

For $C(K, X)$ this has already been asked by E. and P. Saab in [117, Question 12].

An anonymous referee pointed out to us that Hagler and Stegall paper [67] might help to give a solution to this problem. Actually, it was this remark that led us to formulate the solution for $L_\infty(\mu, X)$.

III. All problems considered here can also be posed in more general situations, for instance in locally convex spaces and, in particular in *Fréchet* spaces. This last case is probably the most important. Of course, some work in this direction has already been done. For instance, in connection with J. Hoffmann-Jørgensen and Kwapień's work see [56].

IV. Theorems studied in this monograph have already inspired the search of copies of sequence spaces in several different situations. Let us mention some of them:

(a) Tensor products (see [117]).
(b) Function spaces more general than $L_p(\mu, X)$ spaces, like Orlicz vector valued function spaces $L_\Phi(X)$ (see, for instance [6] or [7]), or even the more general $E(X)$ spaces, where E is an order continuous Banach lattice which has weak unit as considered in [127] (see for instance [53] or [97]).
(c) Vector measure spaces or operator spaces (as in [46], [47] or [121]).
(d) Vector-valued Hardy spaces (as in [44]).

Nevertheless, we believe that in these directions many things remain to be done.

V. It is well known that vector-valued function spaces may be viewed as operator spaces. For example, the space $\mathcal{K}(X, C(K))$ of all compact operators from X into $C(K)$ is isometrically isomorphic to $C(K, X^*)$ [51, VI.7.1], and $L_\infty(\mu, X)$ is very useful in the representation of operators from $L_1(\mu)$ into

X [39, Chapter III]. For this reason the results given here have inspired and have found applications in the study of several operator spaces (see [12], [13], [43], [48], and see also [61] and [24] in connection with [48]). Among these applications are some complementation results more or less related to the famous open problem of the uncomplementability of the space $\mathcal{K}(X, Y)$, of compact operators from X to Y, in the space $\mathcal{L}(X, Y)$ of all operators from X to Y (see [48]). This should not be surprising because it was already shown in [78, Theorem 6] that such a problem is related to the presence of copies of c_0 or ℓ_∞ in operator spaces.

VI. We have so far been mainly interested in *isomorphic* copies of spaces. Another natural point of view is to consider *isometric* copies. Some results in this line have been recently obtained by Koldobsky [81] (see also [65], [103], [108] and [66]).

VII. Instead of asking about subspaces we could also ask about quotients. That is,

Problem 7.5. *When do the spaces considered here have quotients isomorphic to any of the classical sequence spaces?*

In fact Díaz [29] has already got an interesting answer to the preceding question. He has shown that: *"For $1 \le p \le \infty$, if the measure μ is not purely atomic and X^* contains a copy of ℓ_1, then $L_p(\mu, X)$ has a quotient isomorphic to c_0".* This result has allowed him to get a striking characterization of Grothendieck $L_p(\mu, X)$ spaces. The study of quotients was continued by Díaz with Schlüchternann in [34].

The result we have just mentioned recalls us that many properties considered in Banach space theory are connected with the presence of copies or quotients of certain spaces. For instance, property (V^*) of Pełczynski is connected with the existence of complemented ℓ_1-sequences [7], and Grothendieck property with the lack of complemented copies of c_0 [26]. Of course these connections have already been exploited ([7], [25], [26], [29], [59], [104]), but it seems natural to think that they could be used further.

VIII. Let us finally consider the following problem: It is well known that ℓ_∞ and $L_\infty([0, 1])$ are isomorphic [93, p. 111]. However, let us suppose that X has a copy but not a complemented copy of c_0. For instance X may be ℓ_∞ or ℓ_∞/c_0 (a well studied space, see for example [49]), or more generally X may be any Grothendieck $C(K)$ space or a quotient of it. It follows from Theorem 5.1.4 that $L_\infty([0, 1], X)$ *does* contain a complemented copy of c_0, but $\ell_\infty(X)$ *does not*. Hence $L_\infty([0, 1], X)$ and $\ell_\infty(X)$ are not isomorphic. Then the following natural question arises:

Problem 7.6. *In which conditions are $L_\infty([0, 1], X)$ and $\ell_\infty(X)$ isomorphic?*

We do not know the answer even for separable spaces X.

References

1. D. J. Aldous, *Unconditional bases and martingales in $L_p(F)$*, Math. Proc. Camb. Phil. Soc. 85 (1979), 117-123.
2. E.J. Balder, M. Girardi and V. Jalby, *From weak to strong types of $\mathcal{L}^1_E$-convergence by the Bocce criterion*, Studia Math. 111 (1994), 241-263.
3. J. Batt and W. Hiermeyer, *On Compactness in $L_p(\mu, X)$ in the Weak Topology and in the Topology $\sigma(L_p(\mu, X), L_q(\mu, X^*))$*, Math. Z. 182 (1983), 409-423.
4. C. Bessaga and A. Pełczyński, *On bases and unconditional convergence of series in Banach spaces*, Studia Math. 17 (1958), 151-164.
5. F. Bombal, *Algunos Espacios de Banach de Funciones Vectoriales*, Actas de las XII Jornadas Luso-Espanholas de Matemáticas, Vol. I, Universidad do Minho, Braga (1987), 25-60.
6. F. Bombal, *On ℓ^1 subspaces of Orlicz vector-valued function spaces*, Math. Proc. Camb. Phil. Soc. 101 (1987), 107-112.
7. F. Bombal, *On (V^*) sets and Pełczyński's property (V^*)*, Glasgow Math. J. 32 (1990), 109-120.
8. F. Bombal, *Distinguished Subsets in Vector Sequence Spaces*, Progress in Functional Analysis, Proceedings of the Peñíscola Meeting on occasion of 60th birthday of M. Valdivia (Edited by J. Bonet et al.), Elsevier Science Publishers 1992, 293-306.
9. F. Bombal and C. Fierro, *Compacidad débil en espacios de Orlicz de funciones vectoriales*, Rev. Real Acad. Cienc. Exac., Fís. y Natur. Madrid 78 (1984), 157-163.
10. F. Bombal and B. Hernando, *On the injection of a Köthe function space into $L_1(\mu)$*, Annal. Soc. Math. Polon. Serie I: Commentationes Math. 25 (1995), 49-60.
11. F. Bombal and B. Porras, *Strictly Singular and Strictly Cosingular Operators on $C(K, X)$*, Math. Nachr. 143 (1989), 355-364.
12. J. Bonet, P. Domański, M. Lindström and M.S. Ramanujan, *Operator spaces containing c_0 or ℓ_∞*, Results Math. 28 (1995), 250-269.
13. J. Bonet, P. Domański and M. Lindström, *Cotype and complemented copies of c_0 in spaces of operators*, Czechoslovak Math. J. 46 (121) (1996), 271-289.
14. J. Bourgain, *An averaging result for c_0-sequences*, Bull. Soc. Math. Belg. Sér. B 30 (1978), 83-87.
15. J. Bourgain, *A Note on the Lebesgue spaces of vector valued functions*, Bull. Soc. Math. Belg. Sér. B 31 (1979), 45-47.
16. J. Bourgain, *An averaging result for ℓ_1-sequences and applications to weakly conditionally compact sets in $L_1(\mu)$*, Israel J. of Math. 32 (1979), 289-298.
17. J. Bourgain, *The Komlós Theorem for vector valued functions*, unpublished (1979).
18. J. Bourgain, *New classes of $\mathcal{L}^p$-spaces*, Lecture Notes in Math. 889, Springer-Verlag 1981.

19. J. Bourgain and H.P. Rosenthal, *Martingales valued in certain subspaces of L^1*, Israel J. of Math. 37 (1980), 54-75.
20. J.K. Brooks and N. Dinculeanu, *Weak Compactness in Spaces of Bochner Integrable Functions and Applications*, Advances in Math. 24 (1977), 172-188.
21. D. L. Burkholder, *A Geometrical Characterization of Banach Spaces in which Martingale Difference Sequences are Unconditional*, The Annals of Probability 9 (1981), 997-1011.
22. D. L. Burkholder, *Martingales and Fourier Analysis in Banach spaces*, Lecture Notes in Math. 1206: "Probability in Analysis", Springer Verlag (1986), 61-108.
23. D. L. Burkholder, *Explorations in martingale theory and its applications*, Lecture Notes in Math. 1464: "École d'Été de Probabilités de Saint-Flour XIX-1989", Springer Verlag (1991), 1-70.
24. V. Caselles, *A Characterization of Weakly Sequentially Complete Banach Lattices*, Math. Z. 190 (1985), 379-385.
25. P. Cembranos, *Algunas propiedades del espacio de Banach $C(K, X)$*, Ph. D. Thesis, Universidad Complutense de Madrid, Madrid 1984.
26. P. Cembranos, *$C(K, X)$ contains a complemented copy of c_0*, Proc. Amer. Math. Soc. 91 (1984), 556-558.
27. S. Díaz, *Complemented copies of c_0 in $L_\infty(\mu, X)$*, Proc. Amer. Math. Soc. 120 (1994), 1167-1172.
28. S. Díaz, *Complemented copies of ℓ_1 in $L_\infty(\mu, X)$*, preprint.
29. S. Díaz, *Grothendieck's property in $L_p(\mu, X)$*, Glasgow Math. J. 37 (1995), 379-382.
30. S. Díaz, *Weak compactness in $L_1(\mu, X)$*, Proc. Amer. Math. Soc. 124 (1996), 2685-2693.
31. S. Díaz, A. Fernández, M. Florencio and P.J. Paúl, *Complemented copies of c_0 in the space of Pettis integrable functions*, Quaestiones Mathematicae 16 (1993), 61-66.
32. S. Díaz and F. Mayoral, *Oscillation Criteria and Compactenss in Spaces of Bochner Integrable functions*, preprint.
33. S. Díaz and F. Mayoral, *Weak and Norm Compactness in Köthe-Bochner Function Spaces*, preprint.
34. S. Díaz and G. Schlüchternann, *Quotients of vector-valued functions spaces*, preprint.
35. J. Diestel, *Sequences and series in Banach spaces*, Graduate texts in Math. n. 92, Springer-Verlag 1984.
36. J. Diestel, *Uniform Integrability: An Introduction*, Rend. Istit. Mat. Univ. Trieste 23 (1991), 41-80.
37. J. Diestel, H. Jarchow and A. Tonge, *Absolutely Summing Operators*, Cambridge University Press 1995.
38. J. Diestel, W. M. Ruess and W. Schachermayer, *Weak compactness in $L_1(\mu, X)$*, Proc. Amer. Math. Soc. 118 (1993), 447-453.
39. J. Diestel and J. J. Uhl Jr., *Vector Measures*, Math. Surveys n. 15, American Mathematical Society 1977.
40. N. Dinculeanu, *Vector Measures*, Pergamon Press, New York, 1967.
41. J. Dieudonné, *Sur le théorem de Lebesgue-Nikodym (V)*, Canad. J. Math. 3 (1951), 129-139.
42. I. Dobrakov, *On representation of linear operators on $C_0(T, X)$*, Czech. Math. J. 20 (1971), 13-30.
43. P. Domański and L. Drewnowski, *Uncomplementability of the spaces of norm continuous functions in some spaces of "weakly" continuous functions*, Studia Math. 97 (1991), 245-251.

44. P. Dowling, *On complemented copies of c_0 in vector-valued Hardy spaces*, Proc. Amer. Math. Soc. 107 (1989), 251-254.

45. P. Dowling, *A stability property of a class of Banach spaces not containing c_0*, Canadian Math. Bull. 35 (1992), 56-60.

46. L. Drewnowski, *Copies of ℓ_∞ in an operator space*, Math. Proc. Camb. Phil. Soc. 108 (1990), 523-526.

47. L. Drewnowski, *When does $ca(\Sigma, X)$ contain a copy of ℓ_∞ or c_0?*, Proc. Amer. Math. Soc. 109 (1990), 747-752.

48. L. Drewnowski and G. Emmanuele, *The problem of complementability for some spaces of vector measures of bounded variation with values in Banach spaces containing copies of c_0*, Studia Math. 104 (1993), 111-123.

49. L. Drewnowski and J. W. Roberts, *On the primariness of the Banach space ℓ_∞/c_0*, Proc. Amer. Math. Soc. 112 (1991), 949-957.

50. N. Dunford and B.J. Pettis, *Linear operations on summable functions*, Trans. Amer. Math. Soc. 47 (1940), 323-392.

51. N. Dunford and J. T. Schwartz, *Linear Operators*, Part I, Interscience, New York 1958.

52. G. Emmanuele, *On complemented copies of c_0 in $L_p(\mu, X)$, $1 \le p < \infty$* Proc. Amer. Math. Soc. 104 (1988), 785-786.

53. G. Emmanuele, *Copies of ℓ_∞ in Köthe spaces of vector valued functions*, Illinois J. of Math. 36 (1992), 293-296.

54. G. Emmanuele, *Remarks on the existence of copies of c_0 and ℓ_∞ in the space $cabv(\lambda; E)$*, Le Matematiche 50 (1995), 57-63.

55. H. Fakhoury, *Sélections linéaires associées au Théorème de Hahn-Banach*, J. Funct. Anal. 11 (1972), 436-452.

56. X. Fernique, *Sur les spaces de Fréchet ne contenant pas c_0*, Studia Math. 101 (1992), 299-309.

57. J.C. Ferrando, *When does $bvca(\Sigma, X)$ contain a copy of ℓ_∞?*, Math. Scand. 74 (1994), 271-274.

58. J.C. Ferrando, *Copies of c_0 in certain vector-valued function Banach spaces*, Math. Scand. 77 (1995), 148-152.

59. F. Freniche, *Barrelledness of the Space of Vector Valued and Simple Functions*, Math. Ann. 267 (1984), 479-486.

60. F. Freniche, *Embedding c_0 in the space of Pettis integrable functions*, preprint.

61. F. Freniche and L. Rodríguez-Piazza, *Linear projections from a space of measures onto its Bochner integrable functions subspace*, preprint.

62. M. Girardi, *Compactness in L_1, Dundford-Pettis oparators, geometry of Banach spaces*, Proc. Amer. Math. Soc. 111 (1991), 767-777.

63. M. Girardi, *Weak vs. norm compactness in L_1, the Bocce criterion*, Studia Math. 98 (1991), 95-97.

64. G. Godefroy, N.J. Kalton and P.D. Saphar, *Unconditional ideals in Banach spaces*, Studia Math. 104 (1993), 13- 59.

65. P. Greim, *Isometries and L_p-structure of separably valued Bochner L_p-spaces*, Lecture Notes in Math. 1033 (1983), 209-218.

66. S. Guerre and Y. Raynaud, *Sur les isometries de $L^p(X)$ et le thèoréme ergodique vectoriel*, Can. J. Math. 40 (1988), 360-391.

67. J. Hagler and Ch. Stegall, *Banach Spaces Whose Duals Contain Complemented Subspaces Isomorphic to $C([0,1])^*$*, J. Funct. Anal. 13 (1973), 233-251.

68. R. Haydon, *A non-reflexive Grothendieck space that does not contain ℓ_∞*, Israel J. of Math. 40 (1981), 65-73.

69. W. Hensgen, *Some properties of the vector-valued Banach ideal space $E(X)$ derived from those of E and X*, Collectanea Math. 43 (1992), 1-13.

70. W. Hensgen, *A simple proof of Singer's representation theorem,* Proc. Amer. Math. Soc. 124 (1996), 3211-3212.
71. F.L. Hernández and V. Peirats, *A Remark on Sequence F-spaces $\lambda(E)$ Containing a copy of ℓ^p,* Bull. Polish Acad. Sci. Math. 34 (1986), 295-299.
72. E. Hewitt and K. Strongberg, *Real and abstract analysis,* Springer-Verlag, New-York, Heidleberg, Berlin 1965.
73. J. Hoffmann-Jørgensen, *Sums of independent Banach space valued random variables,* Studia Math. 52 (1974), 159-186.
74. Z. Hu and B.-L. Lin, *Extremal Structure of the Unit Ball of $L^p(\mu; X)^*$,* Jour. Math. Anal. Appl. 200 (1996), 567-590.
75. A. and C. Ionescu Tulcea, *Topics in the Theory of Liftings,* Ergebnisse der Mathematik und ihrer Grenzgebiete vol. 48 Springer-Verlag 1969.
76. W.B. Johnson, *A complementary universal conjugate Banach space and its relation to the Approximation Problem,* Israel J. of Math. 13 (1972), 301-310.
77. M.I. Kadec and A. Pełczyński, *Bases, lacunary sequences and complemented subspaces in the spaces L_p,* Studia Math. 21 (1962), 161-176.
78. N. J. Kalton, *Spaces of compact operators,* Math. Ann. 208 (1974), 267-278.
79. N. J. Kalton, *Locally complemented subspaces and $\mathcal{L}_p$-spaces for $0 < p < 1$,* Math. Nachr. 115 (1984), 71-97.
80. A. Koldobsky, *Isometries of $L_p(X; L_q)$ and Equimeasurability,* Indiana Univ. Math. J. 40 (1991), 677-705.
81. A. Koldobsky, *Isometric stability property of certain Banach spaces,* Canadian Math. Bull. 38 (1995), 93-97.
82. J. Kuelbs (Ed.), *Probability on Banach Spaces,* Advances in Probability and Related Topics, Vol. 4, Marcel Dekker 1978.
83. S. Kwapień, *Sur les espaces de Banach contenant c_0. A supplement to the paper by J. Hoffmann-Jørgensen 'Sums of independent Banach space valued random variables',* Studia Math. 52 (1974), 187-188.
84. H. E. Lacey, *The Isometric Theory of Classical Banach Spaces,* Die Grundlehren der mathematischen Wissenschaften in Einzeldarstellungen vol. 208, Springer-Verlag, Berlin Heidelberg New York 1974.
85. S. Lang, *Analysis II,* Addison-Wesley, Reading, MA, 1969.
86. M. Ledoux and M. Talagrand, *Probability in Banach Spaces,* Ergebnisse der Mathematik und ihrer Grenzgebiete 3. Folge, vol. 23, Springer-Verlag 1991.
87. I. E. Leonard, *Banach Sequence Spaces,* Jour. Math. Anal. Appl. 54 (1976), 245-265.
88. D. Leung, *Uniform convergence of operators and Grothendieck spaces with the Dunford-Pettis property,* Thesis, University of Illinois 1987.
89. D. Leung, *Some stability properties of c_0-saturated spaces,* Math. Proc. Camb. Phil. Soc. 118 (1995) 287-301.
90. D. Leung and F. Räbiger, *Complemented copies of c_0 in ℓ_∞-sums of Banach spaces,* Illinois J. of Math. 34 (1990), 52-58.
91. A. Lima, *The metric approximation property, norm-one projections and intersection property of balls,* Israel J. of Math. 84 (1993), 451-475.
92. J. Lindenstrauss and L. Tzafriri, *Classical Banach spaces,* Lecture Notes in Mathematics 338, Springer Verlag 1973.
93. J. Lindenstrauss and L. Tzafriri, *Classical Banach spaces I,* Ergebnisse der Mathematik und ihrer Grenzgebiete vol. 92, Springer-Verlag 1977.
94. J. Lindenstrauss and L. Tzafriri, *Classical Banach spaces II,* Ergebnisse der Mathematik und ihrer Grenzgebicte vol. 97 Springer-Verlag 1979.
95. B. Maurey and G. Pisier, *Séries de variables aléatoires indépendantes et propriétés géométriques des espaces de Banach,* Studia Math. 58 (1976), 45-90.

96. J. Mendoza, *Copies of ℓ_∞ in $L_p(\mu, X)$*, Proc. Amer. Math. Soc. 109 (1990), 125-127.

97. J. Mendoza, *Complemented copies of ℓ_1 in $L_p(\mu, X)$*, Math. Proc. Camb. Phil. Soc. 111 (1992), 531-534.

98. J. Mendoza, *Copies of Classical Sequence Spaces in Vector-Valued Function Banach Spaces*, Function Spaces the Second Conference, Lecture Notes in Pure and Applied Mathematics 172 (edited by K. Jarosz), Marcel Dekker 1995, 311-320.

99. A. Pełczyński, *On Strictly Singular and Strictly Cosingular Operators I*, Bull. Polish Acad. 13 (1965), 31-36.

100. A. Pełczyński and Z. Semadeni, *Spaces of continuous functions (III) (Spaces $C(\Omega)$ for Ω without perfect subsets)*, Studia Math. 18 (1959), 211-222.

101. G. Pisier, *Une propriété de stabilité de la classe des espaces ne contenant pas ℓ_1*, C. R. Acad. Sci. Paris Sér. A 286 (1978),747-749.

102. F. Räbiger, *Beiträge zur Strukturtheorie der Grothendieck-Räume*, in Sitzungsber. Heidelberg Akad. Wiss. Math.-Naturwiss. Klasse, Jg. 1985, 4. Abhandlung, Springer-Verlag, New York 1985.

103. B. Randrianantoanina, *1-Complemented subspaces of spaces with 1-unconditional bases*, preprint.

104. N. Randrianantoanina, *Complemented copies of ℓ_1 and Pełczyński's property (V^*) in Bochner function spaces*, Can. J. Math. 48 (1996), 625-640.

105. N. Randrianantoanina, *Complemented copies of ℓ_1 in spaces of vector measures and applications*, preprint.

106. T.S.S.R.K. Rao, *On locally 1-complemented subspaces of Banach spaces*, preprint.

107. Y. Raynaud, *Sous-espaces ℓ_r et géométrie des espaces $L_p(L_q)$ et L_Φ*, C. R. Acad. Sci. Paris, Sér. I, 301 (1985), 299-302.

108. Y. Raynaud, *Sur les sous-espaces de $L_p(L_q)$*, Seminaire d'Analyse Fonctionelle 1984/85, Publ. Math. Univ. Paris VII, 26, Univ. Paris VII, Paris 1986, 49-71.

109. C. A. Rogers, *Hausdorff Measures*, Cambridge University Press 1970.

110. H. P. Rosenthal, *On relatively disjoint families of measures with some applications to Banach space theory*, Studia Math. 37 (1970), 13-36.

111. H. P. Rosenthal, *On injective Banach spaces and the spaces $L_\infty(\mu)$ for finite measures μ*, Acta Math. 124 (1970), 205-248.

112. H. P. Rosenthal, *Point-wise Compact Subsets of the First Baire Class*, Amer. J. Math. 99 (1977), 362-378.

113. H. P. Rosenthal, *Topics Course, University of Paris VI, 1979*, unpublished.

114. H. P. Rosenthal, *A characterization of Banach spaces containig c_0*, J. of the Amer. Math. Soc. 7 (1994), 707-748.

115. W. Rudin, *Real and Complex Analysis*, Mac Graw-Hill, 1979.

116. E. Saab and P. Saab, *A stability property of a class of Banach spaces not containing a complemented copy of ℓ_1*, Proc. Amer. Math. Soc. 84 (1982), 44-46.

117. E. Saab and P. Saab, *On Stability Problems of Some Properties in Banach Spaces*, Function Spaces, Lecture Notes in Pure and Applied Mathematics 136 (edited by K. Jarosz), Marcel Dekker 1992, 367-394.

118. S. Saks, *Theory of the Integral*, 2nd Edition Hafner, New York 1937 (reimpression: Dover, New York 1964).

119. C. Samuel, *Sur la reproductibilité des espaces ℓ_p*, Math. Scand. 45 (1979), 103-117.

120. C. Samuel, *Sur les sous-espaces de $\ell_p \hat\otimes \ell_q$*, Math. Scand. 47 (1980), 247-250.

121. G. Schlüchtermann, *Properties of Operator-Valued Functions and Applications to Banach Spaces and Linear Semigroups*, Habilitation Thesis. Ludwig Maximilians Universität, München 1994.

122. Th. Schlumprecht, *Limitierte Mengen in Banachräume*, Thesis. Ludwig Maximilians Universität, München 1988.

123. G.L. Seever, *Measures on F-spaces*, Trans. Amer. Math. Soc. 133 (1968), 267-280.

124. Z. Semadeni, *Banach Spaces of Continuous Functions*, PWN-Polish Scientific Publishers, Warszawa 1971.

125. I. Singer, *Best Approximation in Normed Linear Spaces by Elements of Linear Subspaces*, Die Grundlehren der mathematischen Wissenschaften in Einzeldarstellungen vol. 171, Springer-Verlag, Berlin Heidelberg New-York 1970.

126. M. Talagrand, *Un nouveau $C(K)$ qui possède la propriété de Grothendieck*, Israel J. of Math. 37 (1980), 181-191.

127. M. Talagrand, *Weak Cauchy sequences in $L_1(E)$* , Amer. J. of Math. 106 (1984), 703-724.

128. M. Talagrand, *Quand l'espace des measures à variation bornée est-il faiblement séquentiellement complet ?* , Proc. Amer. Math. Soc. 90 (1984), 285-288.

129. A. Ülger, *Weak compactness in $L_1(\mu, X)$*, Proc. Amer. Math. Soc. 113 (1991), 143-149.

Index